GED Math Prep

The Ultimate Step-by-Step Guide Plus
Two Full-Length GED Practice Tests

SCAN ME

Michael Smith
www.mathnotion.com

GED Math Prep

Published By: The Math Notion

Web: www.mathnotion.com

Email: info@mathnotion.com

ISBN: 978-1-63620-024-8

The Math Notion

Michael Smith has been a math instructor for over a decade now. He launched the Math Notion. Since 2006, we have devoted our time to both teaching and developing exceptional math learning materials. As a test prep company, we have worked with thousands of students. We have used the feedback of our students to develop a unique study program that can be used by students to drastically improve their math scores fast and effectively. We have more than a thousand Math learning books including:

– **HiSET Math Prep**
– **TABE Math Prep**
– **TASC Math Prep**
– **Accuplacer Math Prep**
– **Common Core Math Prep**
–**many Math Education Workbooks, Study Guides, Practice and Exercise Books**

As an experienced Math test preparation company, we have helped many students raise their standardized test scores—and attend the colleges of their dreams: We tutor online and in person, we teach students in large groups, and we provide training materials and textbooks through our website and through Amazon.

You can contact us via email at:

info@mathnotion.com

How to Achieve a Perfect Score on the GED Math Test?

GED Math Prep covers all mathematics topics that will be key to succeeding on the GED math test. The step-by-step guide and hundreds of examples in this book can help you hone your math skills, boost your confidence, and be well prepared for the GED test.

This new GED self-teaching prep book offers extensive preparation and brush-up in math for those test takers who plan to take the GED High School Equivalency Test. Two GED practice exams with detailed answers reflect questions and question types found on the actual test.

Updated the new GED prep for 2020, 2021, and beyond by top test prep experts. Inside the GED math preparation, You'll Find:

- Comprehensive GED math review of key concepts

- Content 100% aligned with the latest GED math exam

- Over 2,000 practice questions to help you master each GED math topic

- Lots of examples with step-by-step solutions to illustrate all the GED question types

- Each lesson is short, concise, and to the point.

- Focus on the most challenging part of the GED math test!

- 2 full-length practice exams (featuring new question types) with detailed answers

- And much more than we can fit in this space…

After completing this math preparation book, you will discover your strengths and weaknesses, a strong foundation, and gain confidence to be successful on the GED test. We have helped hundreds of thousands of people pass GED test and achieve their education and career goals. Get math prep for the GED review that you need to ace your exam.

It is an excellent investment in your future!

www.MathNotion.com

... So Much More Online!

✓ FREE Math Lessons

✓ More Math Learning Books!

✓ Mathematics Worksheets

✓ Online Math Tutors

✓ For a PDF Version of This Book

SCAN ME

Please Visit www.mathnotion.com

Contents

Chapter 1 : Whole Numbers, Real Numbers, and Integers...................................**11**

Rounding.. 13

Estimates 15

Whole Number Addition and Subtraction 17

Whole Number Multiplication.. 19

Whole Number Division 21

Adding and Subtracting Integers.. 23

Multiplying and Dividing Integers... 25

Arrange, Order, and Comparing Integers.. 27

Compare Integer 29

Order of Operations 31

Integers and Absolute Value 33

Chapter 2 : Fractions and Decimals..**35**

Simplifying Fractions 37

Factoring Numbers 39

Greatest Common Factor (GCF) 41

Least Common Multiple (LCM) 43

Divisibility Rules 45

Adding and Subtracting Fractions.. 47

Multiplying and Dividing Fractions 49

Adding Mixed Numbers 51

Subtracting Mixed Numbers 53

Multiplying Mixed Numbers 55

Dividing Mixed Numbers 57

Comparing Decimals 59

Rounding Decimals... 61

Adding and Subtracting Decimals 63

Multiplying Decimals 65

Dividing Decimals 67

Converting Between Fractions, Decimals and Mixed Numbers................ 69

Chapter 3 : Proportion, Ratio, Percent **71**

Writing Ratios 73

Simplifying Ratios 75

Create a Proportion.. 77

Similar Figures 79

Ratio and Rates Word Problems 81

Percentage Calculations... 83

Percent Problems 85

Markup, Discount, and Tax 87

Simple Interest 89

Converting Between Percent, Fractions, and Decimals 91

Chapter 4 : Exponents and Radicals **93**

Multiplication Property of Exponents 95

Division Property of Exponents 97

Powers of Products and Quotients... 99

Zero and Negative Exponents 101
Negative Exponents and Negative Bases 103
Writing Scientific Notation 105
Square Roots 107

Chapter 5 : Algebraic Expressions 109
Translate Phrases into an Algebraic Statement 111
The Distributive Property 113
Evaluating One Variable 115
Evaluating Two Variables 117
Expressions and Variables 119
Combining like Terms 121
Expressions 121
Simplifying Polynomial Expressions 123

Chapter 6 : Equations and Inequalities 125
One–Step Equations 127
Two–Step Equations 129
Multi–Step Equations 131
Graphing Single–Variable Inequalities 133
One–Step Inequalities 135
Two–Step Inequalities 137
Multi–Step Inequalities 139
Solving Systems of Equations by Substitution 141
Solving Systems of Equations by Elimination 143
Systems of Equations Word Problems 145
Linear Equations 147
Graphing Lines of Equations 149
Graphing Linear Inequalities 151
Finding Distance of Two Points 153

Chapter 7 : Polynomials 155
Classifying Polynomials 157
Adding and Subtracting Polynomials 159
Multiply and Divide Monomials 161
Multiplying Monomials 163
Multiply a Polynomial and a Monomial 165
Multiply Binomials 167
Factor Trinomials 169
Operations with Polynomials 171
Simplifying Polynomials 173

Chapter 8 : Functions 175
Relations and Functions 177
Rate of change 179
Slope 181
x and y intercept 183
Writing Linear Equations 185
Slope-intercept form 187
Point-slope form 189
Equation of Parallel or Perpendicular lines 191
Equation of Horizontal and Vertical Lines 193
Function Notation 195
Adding and Subtracting Functions 197

Multiplying and Dividing Functions 199
Composition of Functions 201
Solve a Quadratic Equation 203

Chapter 9 : Geometry 205

The Pythagorean Theorem 207
Angles 209
Area of Triangles 211
Area of Trapezoids 213
Area and Perimeter of Polygons 215
Area and Circumference of Circles 217
Volume of Cubes 219
Volume of Rectangle Prisms 221
Surface Area of Cubes 223
Surface Area of a Rectangle Prism 225
Volume of a Cylinder 227
Surface Area of a Cylinder 229

Chapter 10 : Statistics 231

Mean and Median 233
Mode and Range 235
Times Series 237
Box and Whisker Plot 239
Bar Graph 241
Dot plots 243
Scatter Plots 245
Stem–And–Leaf Plot 247
The Pie Graph or Circle Graph 249
Probability of Simple Events 251

Chapter 11 : GED Test Review 253

GED Test Mathematics Formula Sheet 255
GED Practice Test 1 257
GED Practice Test 2 273

Chapter 12 : Answers and Explanations 289

Answer Key 289
Practice Test 1 291
Practice Test 2 298

Chapter 1 : Whole Numbers, Real Numbers, and Integers

Topics that you'll learn in this chapter:

➤ Rounding and Estimates

➤ Addition, Subtraction, Multiplication and Division Whole Number and Integers

➤ Arrange and ordering Integers and Numbers

➤ Comparing Integers, Order of Operations

➤ Mixed Integer Computations

➤ Integers and Absolute Value

"If people do not believe that mathematics is simple, it is only because they do not realize how complicated life is." — John von Neumann

Name: ....

Rounding

Rounding is replacing a number up or down to the closest number or the closest hundred, etc.

- ✓ First, you have to know the place value you'll round to.
- ✓ Second, you have to find the digit to the right of the place value you're rounding to. If it is 5 or greater, add 1 to the place value you're rounding to and put zero for all digits on its right side. If the digit to the right of the place value is smaller than 5 then keep the place value and put zero for all digits to the right.

EXAMPLE:

Round 64 to the closest ten.

The place value of ten is 6. The digit on the right side is 4 (which is smaller than 5). Now keep 6 and put zero for the digit on the right side. Now our answer is 60. 64 is rounded to the closest ten is 60, because 64 is closer to 60 than to 70.

PRACTICES:

Round each number to the underlined place value.

1) 8̲8	2) 8̲.15
3) 4̲,315	4) 56̲5
5) 1.3̲31	6) 14.2̲3
7) 2̲.429	8) 4.31̲3
9) 2.9̲97	10) 7̲.38

Score: ....

Answer Key	
1) 90	2) 8.0
3) 4,000	4) 570
5) 1.3	6) 14.2
7) 2.0	8) 4.31
9) 3.0	10) 7.0

Name: ....

Estimates

Estimating is a math policy used for approximating a number. To estimate *means* to make an irregular guess or calculation. To round means to make easier a known number by scaling it a little bit up or down.

- ✓ To estimate a math problem, round the numbers.
- ✓ For 2-digit numbers, you can usually round to the nearest tens, for 3-digit numbers, round to nearest hundreds, etc.
- ✓ Find the answer.

EXAMPLE:

Estimate the sum by rounding every number to the closest hundred. **153 + 426 =?**

153 is rounded to the closest hundred which is 200. Now 426 is rounded to the closest hundred which is 400.

Then: 200 + 400 = 600

PRACTICES:

Estimate the sum by rounding each added to the nearest ten.

1) 17 + 18	2) 94 + 81
3) 203 + 56	4) 55 + 33
5) 96 + 49	6) 99 + 324
7) 823 + 488	8) 466 + 276
9) 5,112 + 5,792	10) 1,245 + 2,459

Score: ....

Answer Key	
1) 40	2) 200
3) 260	4) 90
5) 150	6) 400
7) 1,300	8) 800
9) 11,000	10) 3,000

Name: ....

Whole Number Addition and Subtraction

- ✓ Arrange the numbers in line.
- ✓ Start with the unit place. (Ones place)
- ✓ Regroup if needed.
- ✓ Add or subtract the tens place.
- ✓ Continue with further digits.

EXAMPLE:

Find the sum. $285 + 145 =$?

First line up the numbers: $\begin{array}{r} 285 \\ +145 \\ \hline \end{array}$ → Start with the unit place. (ones place) $5 + 5 = 10$,

Write 0 for ones place and keep 1, $\begin{array}{r} 1 \\ 285 \\ +145 \\ \hline 0 \end{array}$, Add the tens place and the digit 1 we kept:

$1 + 8 + 4 = 13$, Write 3 and keep 1, $\begin{array}{r} 1\,1 \\ 285 \\ +145 \\ \hline 30 \end{array}$

Continue with further digits → $1 + 2 + 1 = 4$ → $\begin{array}{r} 1\,1 \\ 285 \\ +145 \\ \hline 430 \end{array}$

Find the difference. $976 - 453 =$?

First line up the numbers: $\begin{array}{r} 976 \\ -453 \\ \hline \end{array}$, → Start with the unit place. $6 - 3 = 3$, $\begin{array}{r} 976 \\ -453 \\ \hline 3 \end{array}$,

Subtract the tens place. $7 - 5 = 2$, $\begin{array}{r} 976 \\ -453 \\ \hline 23 \end{array}$, Continue with further digits → $9 - 4 = 5$,

$\begin{array}{r} 976 \\ -453 \\ \hline 523 \end{array}$

PRACTICES:

Find the missing number.

1) 540 − .... = 100	2) 800 − .... = 220
3) .... − 2,650 = 6,700	4) 85,000 − 42,000 = ...
5) 1,280− .... = 420	6) 5,000 + 8,450 = ...
7) .... − 3,870 = 9,630	8) 12,310 − .... = 8,540

Solve.
9) A school had 708 students last year. If all last year students and 218 new students have registered for this year, how many students will there be in total?
10) Lisa had $856 dollars in her saving account. She gave $295 dollars to her brother, Tom. How much money does she have left?

Score: ....

Answer Key	
1) 440	2) 580
3) 9,350	4) 43,000
5) 860	6) 13,450
7) 13,500	8) 3,770
9) 926	10) 561

Name: ....

Whole Number Multiplication

- ✓ First you have to learn the times tables! To solve multiplication problems quick, you need to learn the times table. For example, 3 times 8 is 24 or 8 times 7 is 56.
- ✓ For multiplication, line up the numbers that you are multiplying.
- ✓ Start with the ones place and regroup if needed.
- ✓ Continue with further digits.

EXAMPLE:

Solve. $500 \times 30 = ?$

Line up the numbers: $\begin{array}{r} 500 \\ \times 30 \\ \hline \end{array}$, start with the ones place $\rightarrow 0 \times 500 = 0$, $\begin{array}{r} 500 \\ \times 30 \\ \hline 0 \end{array}$, Continue

with further digit which is 3. $\rightarrow 3 \times 500 = 1,500$, $\begin{array}{r} 500 \\ \times 30 \\ \hline 15,000 \end{array}$

PRACTICES:

Multiply the Number.

1) 120 × 6 = _____	2) 160 × 30 = _____
3) 600 × 30 = _____	4) 420 × 20 = _____
5) 250 × 40 = _____	6) 600 × 40 = _____
7) 215 × 70 = _____	8) 540 × 11 = _____
9) 121 × 10 = _____	10) 254 × 16 = _____

Score: ....

Answer Key	
1) 720	2) 4,800
3) 18,000	4) 8,400
5) 10,000	6) 24,000
7) 15,050	8) 5,940
9) 1,210	10) 4,064

Name: ....

Whole Number Division

- ✓ Division: A typical division problem: Dividend ÷ Divisor = Quotient
- ✓ In division, we want to find how many times a divisor is contained in a dividend. The result we obtain in a division problem is called quotient.
- ✓ First, the problem is written in division format. (Dividend is inside; divisor is outside)

Quotient

Divisor | Dividend

EXAMPLE:

Solve. $234 \div 4 = ?$

First, write the problem in division format.   4 | 234

Start from left digit of the dividend. 4 won't divide 2.

So, we have to choose another digit of the dividend. It is 3.

Now, we will find how many times 4 goes into 23 and the answer is 5.

Write 5 above the dividend part. 4 times 5 is 20.

Write 20 below 23 and subtract. We get the answer 3.

Now take down the next digit which is 4 and find how many times 4 goes into 34?

The answer is 8. Write 8 above dividend.

This is last step since there is no further digit left.

of the dividend to bring down.

The final answer is 58 and we have the remainder 2.

PRACTICES:

Divide the Number.

1) $450 \div 5 =$ _____	2) $320 \div 8 =$ _____
3) $125 \div 25 =$ _____	4) $720 \div 12 =$ _____
5) $588 \div 14 =$ _____	6) $299 \div 13 =$ _____
7) $869 \div 11 =$ _____	8) $801 \div 9 =$ _____
9) $493 \div 17 =$ _____	10) $600 \div 24 =$ _____

Score: ....

Answer Key	
1) 90	2) 40
3) 5	4) 60
5) 42	6) 23
7) 79	8) 89
9) 29	10) 25

Name: ....

Adding and Subtracting Integers

- ✓ Integers include zero, positive natural numbers, and the negative of the natural numbers. $\{\dots, -3, -2, -1, 0, 1, 2, 3, \dots\}$
- ✓ Add a positive integer by putting it to the right on the number line.
- ✓ Add a negative integer by putting it to the left on the number line.
- ✓ Subtract an integer by adding its opposite.

EXAMPLE:

Solve. $(-8) - (-5) =$

We keep the first number and change the sign of the second number to its opposite.

(Change subtraction into addition. Then: $(-8) + 5 = -3$

Solve. $10 + (4 - 8) =$

First subtract the numbers in brackets, $4 - 8 = -4$

Then: $10 + (-4) = \rightarrow$ changes addition into subtraction: $10 - 4 = 6$

PRACTICES:

Find the sum and difference.

1) $8 + (-11)$	2) $(-13) + 25$
3) $(55) - (21)$	4) $(4) - (-5) - (-3)$
5) $2 + (-11) + (-30) + (9)$	6) $(-5) + (-10) + (7 - 19)$
7) $(-20) - (-44)$	8) $(-9) - 13 + 20$
9) $(50) - (-5) + (-25)$	10) $24 + 16 + (-13)$

Score: ....

Answer Key	
1) −3	2) 12
3) 34	4) 12
5) −30	6) −27
7) 24	8) −2
9) 30	10) 27

Name: ....

Multiplying and Dividing Integers

- ✓ (positive) × (positive) = positive
- ✓ (positive) ÷ (positive) = positive
- ✓ (negative) × (negative) = positive
- ✓ (negative) ÷ (negative) = positive
- ✓ (negative) × (positive) = negative
- ✓ (negative) ÷ (positive) = negative
- ✓ (positive) × (negative) = negative
- ✓ (positive) ÷ (negative) = negative

÷ \ ×	+	−
+	+	−
−	−	+

EXAMPLE:

$(+5) \times (+3) = 5 + 5 + 5 = 15$

The basic idea of multiplication is recurrent addition. Example: $5 \times 3 = 5 + 5 + 5 = 15$

We know that division is the inverse operation of multiplication. So, $15 \div 3 = 5$ because $5 \times 3 = 15$ In words, this expression says that 15 may be divided into 3 groups of 5 every because adding five thrice gives 15.

Divide (-91) by (-7)?

Examples on division of integers on different kinds of problems on integers are mentioned here step by step. $(-91) \div (-7) = 13$

PRACTICES:

Find each product and each quotient.

1) $(-8) \times (-5)$	2) $72 \div 9$
3) $4 \times (-5) \times (-6)$	4) $(-95) \div (-5)$
5) $32 \times (-4)$	6) $(-99) \div (-11)$
7) $(-12) \times (-4)$	8) $(-123) \div 1$
9) $(-4) \times (-3) \times 5$	10) $(-0) \div 15$

Score: ....

Answer Key	
1) 40	2) 8
3) 120	4) 19
5) −128	6) 9
7) 48	8) −123
9) 60	10) 0

Name: ....

Arrange, Order, and Comparing Integers

- ✓ When we use a number line, numbers are increased when you go to the right.
- ✓ To compare numbers, you can use number line! As you go from left to right on the number line, you will find a greater number!
- ✓ Order integers from smallest to greatest.

EXAMPLE:

Order integers from to greatest.

$$(-11, -13, 7, -2, 12)$$

To compare numbers, you can use number line! When you see from left to right on the number line, you find a greater number!

$$-13 < -11 < -2 < 7 < 12$$

PRACTICES:

Order each set of integers from least to greatest.	Order each set of integers from greatest to least
1) $2, 6, -15, -11, 1$	2) $1, 17, 6, 8, 65, 2$
3) $9, -8, 3, -2, 11$	4) $-12, 6, -7, 2, -11$
5) $36, -12, 5, 1, -2$	6) $-14, 17, 7, 37, 9$
7) $31, 18, 0, -54, 9, -5$	8) $-54, 0, 14, 19, 15$
9) $-15, -25, -37, 7, 0, 9$	10) $12, 7, -1, -11, 9, -3$

Score: ....

Answer Key	
1) $-15, -11, 1, 2, 6$	2) $65, 17, 8, 6, 2, 1$
3) $-8, -2, 3, 9, 11$	4) $6, 2, -7, -11, -12$
5) $-12, -2, 1, 5, 36$	6) $37, 17, 9, 7, -14$
7) $-54, -5, 0, 9, 18, 31$	8) $19, 15, 14, 0, -54$
9) $-37, -25, -15, 0, 7, 9$	10) $12, 9, 7, -1, -3, -11$

Name: ....

Compare Integer

✓ If you want to compare numbers, you can use a number line! As you move from left to right on the number line, you will find a greater number!

EXAMPLE:

-5 _____ -1

When we compare two integers, we use the symbols < and >.

$-5 < -1$ means that -5 is less than -1

PRACTICES:

Compare. Use >, =, <

1) 4 _____ 3	2) -22 ___ -11
3) 0 _____ -31	4) -41 _____ -12
5) -64 _____ 64	6) -142 _____ -148
7) 68 _____ 100	8) $(-15) \times 6$ _____ $5 \times (-18)$
9) 16 _____ $-(-16)$	10) 405 _____ -405

Score: ....

Answer Key	
1) >	2) <
3) >	4) <
5) <	6) >
7) <	8) =
9) =	10) >

Name: ....

Order of Operations

When you find more than one math operation, use PEMDAS:
- ✓ Parentheses
- ✓ Exponents
- ✓ Multiplication and Division (from left to right)
- ✓ Addition and Subtraction (from left to right)

EXAMPLE:

Solve. $(11 \times 5) - (12 - 25) =$

First you have to simplify inside parentheses: $(11 \times 5) - (12 - 25) = (55) - (-13) =$

Then: $55 + 13 = 68$

PRACTICES:

Evaluate each expression.

1) $24 - (8 \times 6)$	2) $5 \times 6 - (\frac{15}{11 - (-4)})$
3) $12 - (6 \times (-3))$	4) $(6 \times 7) + (-7)$
5) $(\frac{(-1)+4}{(-1)+(-2)}) \times (-9)$	6) $\frac{30}{2(9 - (-1)) - 10}$
7) $58 - (6 \times 9)$	8) $13 + (4 \times 2)$
9) $((-3) + 15) \div (-3)$	10) $[(-8 \div 2) \div (2 - 4))$

Score: ....

Answer Key		
1) −24	2) 29
3) 30	4) 35
5) 9	6) 3
7) 4	8) 21
9) −4	10) 2

Name: ....

Integers and Absolute Value

✓ To find a definite value of a number, simply find its distance from 0 on number line! For example, the distance of 13 and −13 from zero on number line is 13!

EXAMPLE:

Solve. $\frac{|-18|}{9} \times |5 - 8| =$

First find $|-18|$, →the definite value of -18 is 18, then: $|-18| = 18$

$\frac{18}{9} \times |5 - 8| =$

Next, we solve $|5 - 8|$, → $|5 - 8| = |-3|$, the definite value of -3 is 3. $|-3| = 3$

Then: $\frac{18}{9} \times 3 = 2 \times 3 = 6$

PRACTICES:

Write absolute value of each number.

1) 62	2) − 32
3) − 11	4) 5

Evaluate.

5) $|-12| - |3| + 2$	6) $19 + |-5 - 14| - |2|$
7) $|-11| + |-9|$	8) $|91| - |-18| - 18$
9) $|-10 + 4| \times \frac{|-7 \times 5|}{7}$	10) $\frac{|-16 \times 3|}{2} \times |-12|$

Score: ....

Answer Key	
1) 62	2) 32
3) 11	4) 5
5) 11	6) 36
7) 20	8) 55
9) 30	10) 288

Chapter 2 : Fractions and Decimals

Topics that you'll learn in this chapter:

➢ Simplifying Fractions

➢ Adding and Subtracting Fractions, Mixed Numbers and Decimals

➢ Multiplying and Dividing Fractions, Mixed Numbers and Decimals

➢ Comparing and Rounding Decimals

➢ Converting Between Fractions, Decimals and Mixed Numbers

➢ Factoring Numbers, Greatest Common Factor, and Least Common Multiple

➢ Divisibility Rules

"A Man is like a fraction whose numerator is what he is and whose denominator is what he thinks of himself. The larger the denominator, the smaller the fraction." –Tolstoy

Name: ....

Simplifying Fractions

✓ Regularly divide both the top and bottom of the fraction by $2, 3, 5, 7, \ldots$ etc.
✓ Continue until you can't go any further.

EXAMPLE:

Simplify $\dfrac{12}{20}$.

To simplify $\dfrac{12}{20}$, you have to find a number that both 12 and 20 are divisible by. Both are divisible by 4. Then: $\dfrac{12}{20} = \dfrac{12 \div 4}{20 \div 4} = \dfrac{3}{5}$

PRACTICES:

Simplify the fractions.

1) $\dfrac{44}{64}$

2) $\dfrac{12}{26}$

3) $\dfrac{15}{25}$

4) $\dfrac{30}{45}$

5) $\dfrac{18}{27}$

6) $1\dfrac{62}{124}$

7) $4\dfrac{12}{66}$

8) $1\dfrac{55}{70}$

9) $\dfrac{54}{60}$

10) $7\dfrac{68}{136}$

Score: ....

<table>
<tr><td colspan="2" align="center">Answer Key</td></tr>
<tr><td>1) $\frac{11}{16}$</td><td>2) $\frac{6}{13}$</td></tr>
<tr><td>3) $\frac{3}{5}$</td><td>4) $\frac{2}{3}$</td></tr>
<tr><td>5) $\frac{2}{3}$</td><td>6) $1\frac{1}{2}$</td></tr>
<tr><td>7) $4\frac{2}{11}$</td><td>8) $1\frac{11}{14}$</td></tr>
<tr><td>9) $\frac{9}{10}$</td><td>10) $7\frac{1}{2}$</td></tr>
</table>

Name: ....

Factoring Numbers

- ✓ To break the numbers into their prime factors is called factoring.

- ✓ First few prime numbers are $2, 3, 5, 7, 11, 13, 17, 19$

EXAMPLE:

List all positive factors of 12.

Write the upside-down division:

The second column is the answer.

12	2
6	2
3	3
1	

Then: $12 = 2 \times 2 \times 3$ or $12 = 2^2 \times 3$

PRACTICES:

List all positive factors of each number.	List the prime factorization for each number.
1) 90	2) 40
3) 49	4) 105
5) 50	6) 42
7) 34	8) 78
9) 96	10) 165

Score: ....

Answer Key	
1) $1, 2, 3, 5, 6, 9, 10, 15, 18, 30, 45, 90$	2) $2 \times 2 \times 2 \times 5$
3) $1, 7, 49$	4) $3 \times 5 \times 7$
5) $1, 2, 5, 10, 25, 50$	6) $2 \times 3 \times 7$
7) $1, 2, 17, 34$	8) $2 \times 3 \times 13$
9) $1, 2, 3, 4, 6, 8, 12, 16, 24, 32, 48, 96$	10) $3 \times 5 \times 11$

Name: ....

Greatest Common Factor (GCF)

✓ List the prime factors of each number.

✓ Then multiply common prime factors.

✓ If there are no common prime factors, then our GCF is 1.

EXAMPLE:

Find the GCF for **10** and **15**.

The factors of 10 are: $\{1, 2, 5, 10\}$

The factors of 15 are: $\{1, 3, 5, 15\}$

There is 5 in common,

Then the greatest common factor is: 5

PRACTICES:

Find the GCF for each number pair.

1) 12, 25	2) 72, 84
3) 24, 36	4) 30, 45
5) 9, 36	6) 63, 42
7) 27, 12	8) 125, 50
9) 54, 39	10) 36, 52

Score: ....

Answer Key	
1) 1	2) 12
3) 12	4) 15
5) 9	6) 21
7) 3	8) 25
9) 3	10) 4

Name: ....

Least Common Multiple (LCM)

✓ The smallest multiple that 2 or more numbers have in common is called least common multiple of that number. How to find LCM:

✓ First find the list of the prime factors of each number.

✓ Then multiply the common prime factors and uncommon prime factors of the numbers (each common prime factor is used only for once)

EXAMPLE:

Find the LCM for **18** and **12**.

Prime Numbers: 2 x 2 x 3   2 x 3 x 3

Common factors: 2 and 3
Not Common: 2 and 3

2 x 2 x 3 x 3 = 36

PRACTICES:

Find the LCM for each number pair.

1) 12, 9	2) 40, 20
3) 15, 30	4) 84, 60
5) 60, 40	6) 52, 78
7) 14, 28	8) 14, 7, 42
9) 24, 32	10) 72, 66, 24

Score: ....

Answer Key

1) 36	2) 40
3) 30	4) 420
5) 120	6) 156
7) 28	8) 42
9) 96	10) 792

Name: ....

Divisibility Rules

If a number can be divided by other numbers, it is referred as divisibility. The number is divisible:

- ✓ by 2 if the number is found even.
- ✓ by 3 if the sum of the digits is found to be divisible by 3.
- ✓ by 9 if the sum of the digits is found to be divisible by 9.
- ✓ by 4, if the last 2 digits of a number are found to be divisible by 4.
- ✓ by 6, if it is found to be divisible by 2 and 3.
- ✓ by 8, if it is found to be divisible by 2 and 4.
- ✓ by 5 if the last digit is found 0 or 5.
- ✓ by 10 if the last digit is 0.

EXAMPLE:

What is the factor of 240?

2, because the number is even   3, because $(2 + 4 + 0 = 6, 6 \div 3 = 2)$

4, because $(40 \div 4 = 10)$ 5, because the last digit is 0

8, because of 2 and 4   10, because the last digit is 0

Then 240 is divisible by 2, 3, 4, 5, 8, 10

PRACTICES:

Use the divisibility rules to find the factors of each number.

1) 12	2 3 4 5 6 7 8 9 10	2) 24	2 3 4 5 6 7 8 9 10
3) 36	2 3 4 5 6 7 8 9 10	4) 18	2 3 4 5 6 7 8 9 10
5) 30	2 3 4 5 6 7 8 9 10	6) 54	2 3 4 5 6 7 8 9 10
7) 90	2 3 4 5 6 7 8 9 10	8) 80	2 3 4 5 6 7 8 9 10
9) 72	2 3 4 5 6 7 8 9 10	10) 84	2 3 4 5 6 7 8 9 10

Score: ....

Answer Key

1) 12	<u>2</u> <u>3</u> <u>4</u> 5 <u>6</u> 7 8 9 10	2) 24	<u>2</u> <u>3</u> <u>4</u> 5 <u>6</u> 7 <u>8</u> 9 10
3) 36	<u>2</u> <u>3</u> <u>4</u> 5 <u>6</u> 7 8 <u>9</u> 10	4) 18	<u>2</u> <u>3</u> 4 5 <u>6</u> 7 8 <u>9</u> 10
5) 30	<u>2</u> <u>3</u> 4 <u>5</u> <u>6</u> 7 8 9 <u>10</u>	6) 54	<u>2</u> <u>3</u> 4 5 <u>6</u> 7 8 <u>9</u> 10
7) 90	<u>2</u> <u>3</u> 4 <u>5</u> <u>6</u> 7 8 <u>9</u> <u>10</u>	8) 80	<u>2</u> 3 <u>4</u> <u>5</u> 6 7 <u>8</u> 9 <u>10</u>
9) 72	<u>2</u> <u>3</u> <u>4</u> 5 <u>6</u> 7 <u>8</u> <u>9</u> 10	10) 84	<u>2</u> <u>3</u> <u>4</u> 5 <u>6</u> <u>7</u> 8 9 10

Name: ....

Adding and Subtracting Fractions

✓ Find equivalent fractions with the equivalent divisor before you can add or subtract fractions with totally different divisors.

✓ Adding and Subtracting with the equivalent divisors:

$$\frac{a}{b} + \frac{c}{b} = \frac{a+c}{b} \, , \frac{a}{b} - \frac{c}{b} = \frac{a-c}{b}$$

✓ Adding and Subtracting fractions with different divisors:

$$\frac{a}{b} + \frac{c}{d} = \frac{ad+cb}{bd} \, , \frac{a}{b} - \frac{c}{d} = \frac{ad-cb}{bd}$$

EXAMPLE:

Subtract fractions. $\frac{2}{3} - \frac{1}{2} = ?$

For "unlike" fractions, find equivalent fractions with the same divisors before you can add or subtract fractions with different divisors. Use this formula: $\frac{a}{b} - \frac{c}{d} = \frac{ad-cb}{bd}$

$$\frac{2}{3} - \frac{1}{2} = \frac{(2)(2) - (1)(3)}{3 \times 2} = \frac{4-3}{6} = \frac{1}{6}$$

PRACTICES:

Add fractions.	Subtract fractions.
1) $\frac{1}{4} + \frac{2}{3}$	2) $\frac{1}{2} - \frac{1}{5}$
3) $\frac{1}{3} + \frac{1}{2}$	4) $\frac{1}{7} - \frac{1}{9}$
5) $\frac{1}{4} + \frac{5}{7}$	6) $\frac{3}{5} - \frac{1}{15}$
7) $\frac{6}{7} + \frac{3}{21}$	8) $\frac{1}{3} - \frac{1}{4}$
9) $\frac{5}{13} + \frac{1}{2}$	10) $\frac{6}{5} - \frac{5}{6}$

Score: ....

Answer Key

1) $\frac{11}{12}$	2) $\frac{3}{10}$
3) $\frac{5}{6}$	4) $\frac{2}{63}$
5) $\frac{27}{28}$	6) $\frac{8}{15}$
7) 1	8) $\frac{1}{12}$
9) $\frac{23}{26}$	10) $\frac{11}{30}$

Name: ....

Multiplying and Dividing Fractions

- ✓ How to multiply fractions: multiply the top numbers and multiply the bottom numbers.
- ✓ How to change fractions: Keep, Change, Flip
- ✓ Keep the first fraction then change division sign into multiplication sign and flip the numerator and denominator of the second fraction. Then, solve it

EXAMPLE:

Multiplying fractions. $\frac{5}{6} \times \frac{3}{4} =$

Multiply the upper numbers and multiply the lower numbers.

$\frac{5}{6} \times \frac{3}{4} = \frac{5 \times 3}{6 \times 4} = \frac{15}{24}$, simplify: $\frac{15}{24} = \frac{15 \div 3}{24 \div 3} = \frac{5}{8}$

Dividing fractions. $\frac{1}{4} \div \frac{2}{3} =$

Keep the first fraction then change division sign into multiplication sign and flip the numerator and denominator of the second fraction. Then: $\frac{1}{4} \times \frac{3}{2} = \frac{1 \times 3}{4 \times 2} = \frac{3}{8}$

PRACTICES:

Multiplying fractions. Then simplify.	Dividing fractions.
1) $\frac{3}{5} \times \frac{5}{9}$	2) $\frac{4}{9} \div 4$
3) $\frac{5}{21} \times \frac{7}{10}$	4) $\frac{32}{25} \div \frac{8}{5}$
5) $\frac{3}{29} \times \frac{29}{3}$	6) $\frac{2}{7} \div \frac{8}{35}$
7) $\frac{8}{11} \times 11$	8) $\frac{12}{25} \div \frac{3}{5}$
9) $\frac{7}{9} \times \frac{12}{28}$	10) $7 \div \frac{2}{3}$

Score: ....

Answer Key	
1) $\frac{1}{3}$	2) $\frac{1}{9}$
3) $\frac{1}{6}$	4) $\frac{4}{5}$
5) 1	6) $1\frac{1}{4}$
7) 8	8) $\frac{4}{5}$
9) $\frac{1}{3}$	10) $10\frac{1}{2}$

Name: ....

Adding Mixed Numbers

Use these steps for both adding and subtracting mixed numbers.
- ✓ Add whole number of the mixed numbers.
- ✓ Add the fractions of every mixed number.
- ✓ Find the Least Common Divisor (LCD) if needed.
- ✓ Add whole numbers and fractions.
- ✓ Write your answer in simplest form.

EXAMPLE:

$1\frac{3}{4} + 2\frac{3}{8} = ?$

Rewriting our equation with parts separated, $1 + \frac{3}{4} + 2 + \frac{3}{8}$, Solving the whole number

parts $1 + 2 = 3$, Solving the fraction parts $\frac{3}{4} + \frac{3}{8}$, and rewrite to solve with the equivalent

fractions.

$\frac{6}{8} + \frac{3}{8} = \frac{9}{8} = 1\frac{1}{8}$, then combining the whole and fraction parts $3 + 1 + \frac{1}{8} = 4\frac{1}{8}$

PRACTICES:

Add.

1) $1\frac{1}{7} + 2\frac{1}{3}$	2) $1\frac{1}{2} + 3\frac{2}{3}$
3) $1\frac{2}{5} + 2\frac{1}{10}$	4) $7 + 2\frac{1}{2}$
5) $4\frac{1}{3} + 2\frac{2}{3}$	6) $2\frac{2}{3} + 1\frac{1}{4}$
7) $2\frac{3}{4} + 3\frac{1}{8}$	8) $9 + 1\frac{1}{9}$
9) $4\frac{5}{12} + 2\frac{3}{4}$	10) $3\frac{1}{7} + 2\frac{3}{14}$

Score: ....

Answer Key	
1) $3\frac{10}{21}$	2) $5\frac{1}{6}$
3) $3\frac{1}{2}$	4) $9\frac{1}{2}$
5) 7	6) $3\frac{11}{12}$
7) $5\frac{7}{8}$	8) $10\frac{1}{9}$
9) $7\frac{1}{6}$	10) $5\frac{5}{14}$

Name: ....

Subtracting Mixed Numbers

Use these steps for both adding and subtracting mixed numbers.
- ✓ From whole number of the first mixed number, subtract the whole number of second mixed number.
- ✓ From first fraction subtract the second.
- ✓ Find the Least Common Divisor (LCD) if needed.
- ✓ Add the result of whole numbers and fractions.
- ✓ Write your answer in simplest terms.

EXAMPLE:

$5\frac{2}{3} - 2\frac{1}{4} = ?$

Rewriting our equation with parts separated, $5 + \frac{2}{3} - 2 - \frac{1}{4}$

Solving the whole number parts $5 - 2 = 3$, Solving the fraction parts, $\frac{2}{3} - \frac{1}{4} = \frac{8-3}{12} = \frac{5}{12}$

Joining the whole and fraction parts, $3 + \frac{5}{12} = 3\frac{5}{12}$

PRACTICES:

Subtract.

1) $5\frac{2}{7} - 2\frac{1}{14}$

2) $4\frac{2}{5} - \frac{2}{3}$

3) $3\frac{3}{7} - 1\frac{1}{14}$

4) $7\frac{2}{5} - 5\frac{1}{3}$

5) $4\frac{1}{2} - 2\frac{4}{8}$

6) $11\frac{5}{12} - 8\frac{3}{4}$

7) $7\frac{5}{12} - 5\frac{7}{12}$

8) $5\frac{2}{9} - 2\frac{1}{18}$

9) $3\frac{2}{5} - 2\frac{1}{5}$

10) $3\frac{4}{9} - 1\frac{2}{9}$

Score: ....

Answer Key	
1) $3\frac{3}{14}$	2) $3\frac{11}{15}$
3) $2\frac{5}{14}$	4) $2\frac{1}{15}$
5) 2	6) $2\frac{2}{3}$
7) $1\frac{5}{6}$	8) $3\frac{1}{6}$
9) $1\frac{1}{5}$	10) $2\frac{2}{9}$

Name: ....

Multiplying Mixed Numbers

✓ Convert the mixed numbers into improper fractions. (Improper fraction is a fraction in which the numerator is greater than denominator)

✓ Multiply fractions and write in simplest form if needed.

$$a\frac{c}{b} = a + \frac{c}{b} = \frac{ab + c}{b}$$

EXAMPLE:

Multiply mixed numbers. $4\frac{3}{5} \times 2\frac{1}{3} = ?$

Changing mixed numbers to fractions, $\frac{23}{5} \times \frac{7}{3}$, Applying the fractions formula for

multiplication, $\frac{23 \times 7}{5 \times 3} = \frac{161}{15} = 10\frac{11}{15}$

PRACTICES:

Find each product.

1) $2\frac{1}{3} \times \frac{1}{2}$	2) $1\frac{2}{5} \times \frac{2}{3}$
3) $2\frac{4}{3} \times 2\frac{2}{6}$	4) $2\frac{1}{2} \times 1\frac{2}{4}$
5) $3\frac{1}{2} \times 1\frac{2}{3}$	6) $1\frac{1}{7} \times 1\frac{3}{4}$
7) $1\frac{1}{4} \times 2\frac{6}{5}$	8) $3\frac{1}{2} \times 4\frac{2}{5}$
9) $1\frac{2}{5} \times 2\frac{1}{3}$	10) $5\frac{7}{12} \times 2\frac{4}{9}$

Score: ....

	Answer Key	
1) $1\frac{1}{6}$	2) $\frac{14}{15}$
3) $7\frac{7}{9}$	4) $3\frac{3}{4}$
5) $5\frac{5}{6}$	6) 2
7) 4	8) $15\frac{2}{5}$
9) $3\frac{4}{15}$	10) $13\frac{35}{54}$

Name: ....

Dividing Mixed Numbers

- ✓ Change the mixed numbers into improper fractions.

- ✓ Divide fractions and write in simplest form if needed.

$$a\frac{c}{b} = a + \frac{c}{b} = \frac{ab + c}{b}$$

EXAMPLE:

Find the quotient. $2\frac{1}{2} \div 1\frac{1}{5} = ?$

Changing mixed numbers to fractions, $\frac{5}{2} \div \frac{6}{5}$, Applying the fractions formula for

multiplication, $\frac{5}{2} \times \frac{5}{6} = \frac{5 \times 5}{2 \times 6} = \frac{25}{12} = 2\frac{1}{12}$

PRACTICES:

Find each quotient.

1) $2\frac{3}{5} \div 1\frac{3}{8}$	2) $\frac{3}{2} \div 2\frac{3}{4}$
3) $1\frac{4}{7} \div 2\frac{2}{3}$	4) $1\frac{2}{3} \div 2\frac{1}{3}$
5) $0 \div 4\frac{2}{5}$	6) $2\frac{2}{5} \div 1\frac{1}{2}$
7) $1\frac{2}{3} \div 2\frac{1}{5}$	8) $3\frac{2}{7} \div 4\frac{3}{5}$
9) $1\frac{1}{4} \div 2\frac{4}{5}$	10) $2 \div 3\frac{1}{3}$

Score: ....

Answer Key	
1) $1\frac{49}{55}$	2) $\frac{6}{11}$
3) $\frac{33}{56}$	4) $\frac{5}{7}$
5) 0	6) $1\frac{3}{5}$
7) $\frac{25}{33}$	8) $\frac{5}{7}$
9) $\frac{25}{56}$	10) $\frac{3}{5}$

Name: ....

Comparing Decimals

Decimal is a fraction written in a unique form. For example, instead of writing $\frac{1}{2}$ you can write as **0.5**.

For comparison of decimals:

✓ Compare every digit of two decimals in the same place value.

✓ Start from left. Compare ones, tens, hundreds, tenths, hundredths, etc.

✓ To compare numbers, use these symbols:

- Equal to =, Less than <, Greater than >

- Greater than or equal ≥, Less than or equal ≤

EXAMPLE:

Compare **0.20** and **0.02**.

0.20 *is greater than* 0.02, because the tenth place of 0.20 is 2, but the tenth place of 0.02 is zero. Then: 0.20 > 0.02

PRACTICES:

Write the correct comparison symbol (>, < or =).

1) 0.025 _____ 0.25	2) 0.9 _____ 0.888
3) 4.510 _____ 4.150	4) 10.01 _____ 10.10
5) 0.987 _____ 0.991	6) 18.004 _____ 18.040
7) 0.020 _____ 0.20	8) 0.071___ _0.700
9) 0.08____0.009	10) 0.690 _____ 0.609

Score: ....

Answer Key	
1) <	2) >
3) >	4) <
5) <	6) <
7) <	8) <
9) >	10) >

Name: ....

Rounding Decimals

- ✓ To round a decimal, you must find the place value you'll round to.
- ✓ Then, find the digit to the right of the place value you're rounding to.
  - If it is 5 or greater, add 1 to the place value you're rounding to and remove all digits on its right side.
  - If the digit to the right of the place value is smaller than 5, keep the place value and remove all digits on the right.

EXAMPLE:

Round 2.1837 to the thousandth-place value.

First have a look at the next place value to the right, (tens thousandths). It's 7 and it's found to be greater than 5. So, add 1 to the digit in the thousandth place.

Thousandth place is 3. → 3 + 1 = 4, then, the answer is 2.184

PRACTICES:

Round each decimal number to the nearest place indicated.

1) 6.0<u>8</u>	2) 12.2<u>6</u>7
3) 9.<u>3</u>01	4) 10.0<u>7</u>1
5) 5<u>5</u>.89	6) 59.<u>1</u>5
7) 3<u>2</u>9.018	8) 92.<u>4</u>10
9) 1.4<u>9</u>9	10) 2<u>5</u>.621

Score: ....

Answer Key	
1) 6.1	2) 12.3
3) 9.3	4) 10.07
5) 56	6) 59
7) 330	8) 92.4
9) 1.5	10) 26

Name: ....

Adding and Subtracting Decimals

- ✓ Arrange the numbers in line.
- ✓ Add zeros to have same number of digits for both the numbers.
- ✓ Add or subtract by using column subtraction or addition.

EXAMPLE:

Add. $2.5 + 1.24 =$

First line up the numbers: $\begin{array}{r} 2.5 \\ +1.24 \\ \hline \end{array}$ → Add zeros to have same number of digits for both

numbers. $\begin{array}{r} 2.50 \\ +1.24 \\ \hline \end{array}$, Start with the hundredths place. $0+4=4$, $\begin{array}{r} 2.50 \\ +1.24 \\ \hline 4 \end{array}$, Continue with tenths

place. $5+2=7$, $\begin{array}{r} 2.50 \\ +1.24 \\ \hline .74 \end{array}$. Add the ones place. $2+1=3$, $\begin{array}{r} 2.50 \\ +1.24 \\ \hline 3.74 \end{array}$

Subtract decimals. $4.67 - 2.15 = \begin{array}{r} 4.67 \\ -2.15 \\ \hline \end{array}$

Start with the hundredths place. $7-5=2$, $\begin{array}{r} 4.67 \\ -2.15 \\ \hline 2 \end{array}$, continue with tenths place. $6-1=5$

$\begin{array}{r} 4.67 \\ -2.15 \\ \hline .52 \end{array}$, subtract the ones place. $4-2=2$, $\begin{array}{r} 4.67 \\ -2.15 \\ \hline 2.52 \end{array}$.

PRACTICES:

Add and subtract decimals.

1) $\begin{array}{r} 87.15 \\ -\ 32.35 \\ \hline \end{array}$	2) $\begin{array}{r} 90.43 \\ +\ 44.09 \\ \hline \end{array}$
3) $\begin{array}{r} 58.56 \\ +\ 12.10 \\ \hline \end{array}$	4) $\begin{array}{r} 65.23 \\ -\ 56.48 \\ \hline \end{array}$
5) $\begin{array}{r} 98.125 \\ +\ 58.54 \\ \hline \end{array}$	6) $\begin{array}{r} 162.05 \\ -\ 83.65 \\ \hline \end{array}$

Solve.

7) $\_\_ + 5.0 = 9.08$	8) $7.06 + \_\_ = 24.6$
9) $21.9 - \_\_ = 6.9$	10) $32.12 - \_\_ = 12.07$

Score: ....

Answer Key	
1) 54.8	2) 134.52
3) 70.66	4) 8.75
5) 156.665	6) 78.4
7) 4.08	8) 17.54
9) 15	10) 20.05

Name: ....

Multiplying Decimals

- ✓ Arrange and multiply the numbers as you do with whole numbers.
- ✓ Then count the total number of decimal places in every factor.
- ✓ Place the decimal point in the product.

EXAMPLE:

Find the product. $0.50 \times 0.20 =$

Arrange and multiply the numbers as you do with whole numbers. Line up the numbers:

$\begin{array}{r} 50 \\ \times 20 \\ \hline \end{array}$, Start with the ones place $\rightarrow 0 \times 50 = 0$, $\begin{array}{r} 50 \\ \times 20 \\ \hline 0 \end{array}$, Continue with other digits $\rightarrow$

$2 \times 50 = 100$, $\begin{array}{r} 50 \\ \times 20 \\ \hline 1{,}000 \end{array}$, Count the total number of decimal places in both of the factors

(4). Then Place the decimal point in the product.

Then: $\begin{array}{r} 0.50 \\ \times 0.20 \\ \hline 0.1000 \end{array}$ $\rightarrow 0.50 \times 0.20 = 0.1$

PRACTICES:

Find each product.

1) $\begin{array}{r} 1.5 \\ \times\,0.16 \\ \hline \end{array}$	2) $\begin{array}{r} 5.3 \\ \times\,1.9 \\ \hline \end{array}$
3) $\begin{array}{r} 0.06 \\ \times\,2.5 \\ \hline \end{array}$	4) $\begin{array}{r} 3.19 \\ \times\,21.5 \\ \hline \end{array}$
5) $\begin{array}{r} 9.3 \\ \times\,11.5 \\ \hline \end{array}$	6) $\begin{array}{r} 3.01 \\ \times\,2.1 \\ \hline \end{array}$
7) $\begin{array}{r} 5.0 \\ \times\,1.4 \\ \hline \end{array}$	8) $\begin{array}{r} 23.8 \\ \times\,10 \\ \hline \end{array}$
9) $\begin{array}{r} 21.5 \\ \times\,0.001 \\ \hline \end{array}$	10) $\begin{array}{r} 8.21 \\ \times\,3.1 \\ \hline \end{array}$

Score: ....

Answer Key	
1) 0.24	2) 10.07
3) 0.15	4) 68.585
5) 106.95	6) 6.321
7) 7	8) 238
9) 0.0215	10) 25.451

Name: ....

Dividing Decimals

- ✓ If the divisor is not a whole number, transfer decimal point to right to make it a whole number. Do the same step for dividend.
- ✓ Divide same to whole numbers.

EXAMPLE:

Find the quotient. $1.20 \div 0.2 =$

The divisor is not a whole number. Multiply it by 10 to get 2. Do the same step for the dividend to get 12. Now, divide: $12 \div 2 = 6$. The answer is 6.

PRACTICES:

Find each quotient.

1) $25.7 \div 0.5$	2) $67.2 \div 4$
3) $61.75 \div 1.9$	4) $18.0 \div 1.2$
5) $12.4 \div 10$	6) $2.2 \div 100$
7) $1.88 \div 100$	8) $55.1 \div 100$
9) $0.1 \div 100$	10) $0.25 \div 10$

Score: ....

Answer Key	
1) 51.4	2) 16.8
3) 32.5	4) 15
5) 1.24	6) 0.022
7) 0.0188	8) 0.551
9) 0.001	10) 0.025

Name: ....

Converting Between Fractions, Decimals and Mixed

How to convert fraction into Decimal:

✓ Divide the numerator by denominator.

How to convert decimal into Fraction:

✓ Write decimal over 1.

✓ Multiply both numerator value and denominator value by 10 for each digit on the right side of the decimal point.

✓ Make it to simplest form.

EXAMPLE:

What is long division of $\frac{5}{8}$ =?

In that case we put extra zeros and did $\frac{5.000}{8}$ to get 0.625

PRACTICES:

Convert fractions to decimals.	Convert decimal into fraction or mixed numbers
1) $\frac{4}{10}$	2) 3.6
3) $\frac{3}{8}$	4) 0.07
5) $\frac{4}{12}$	6) 0.15
7) $\frac{5}{16}$	8) 2.7
9) $\frac{60}{100}$	10) 2.5

Score: ....

Answer Key	
1) 0.4	2) $3\frac{3}{5}$
3) 0.375	4) $\frac{7}{100}$
5) 0.333	6) $\frac{3}{20}$
7) 0.3125	8) $2\frac{7}{10}$
9) 0.6	10) $2\frac{1}{2}$

Chapter 3 : Proportion, Ratio, Percent

Topics that you'll learn in this chapter:

- ➢ Writing and Simplifying Ratios

- ➢ Create a Proportion

- ➢ Similar Figures

- ➢ Simple Interest

- ➢ Ratio and Rates Word Problems

- ➢ Percentage Calculations

- ➢ Converting Between Percent, Fractions, and Decimals

- ➢ Percent Problems

- ➢ Markup, Discount, and Tax

"Do not worry about your difficulties in mathematics. I can assure you mine are still greater." – *Albert Einstein*

Name: ....

Writing Ratios

✓ A ratio is a comparison of two numbers, and it can be written as a division.

EXAMPLE:

$3 : 5 = ?$

Both numbers 3 and 5 are divisible by 8 , $\Rightarrow 3 \div 8 = \frac{3}{8}, 5 \div 8 = \frac{5}{8}$,

Then: $3 : 5 = \frac{3}{8}$ and $\frac{5}{8}$.

PRACTICES:

Express each ratio as a rate and unite rate.	Express each ratio as a fraction in the simplest form
1) 80 dollars for 4 chairs.	2) 13 cups to 39 cups.
3) 125 miles on 25 gallons of gas.	4) 17 cakes out of 51 cakes
5) 147 miles on 7 hours.	6) 35 red desks out of 125 desks
7) 12 inches of snow in 24 hours.	8) 8 story books out of 32 books
9) 14 dimes to 112 dimes.	10) 12 gallons to 20 gallons

Score: ....

Answer Key

1)	$\frac{80 \text{ dollars}}{4 \text{ books}}$, 20.00 dollars per chair	2)	$\frac{1}{3}$
3)	$\frac{125 \text{ miles}}{25 \text{ gallons}}$, 5 miles per gallon	4)	$\frac{1}{3}$
5)	$\frac{147 \text{ miles}}{7 \text{ hours}}$, 21 miles per hour	6)	$\frac{7}{25}$
7)	$\frac{12" \text{ of snow}}{24 \text{ hours}}$, 0.5 inches of snow per hour	8)	$\frac{1}{4}$
9)	$\frac{14 \text{ dimes}}{112 \text{ dimes}}$, $\frac{1}{8}$ per dime	10)	$\frac{3}{5}$

Name: ....

Simplifying Ratios

- ✓ Ratios are used to compare two numbers.

- ✓ Ratios can be written as a fraction, using colon or the word "to".

- ✓ You can calculate identical ratios by multiplying or dividing both sides of the ratio by the same number.

EXAMPLE:

Simplify. $8:4 =$

Both numbers 8 and 4 are divisible by 4 , $\Rightarrow 8 \div 4 = 2, 4 \div 4 = 1,$

Then: $8:4 = 2:1$

PRACTICES:

Reduce each ratio.

1) 49: 14	2) 22: 55
3) 35: 25	4) 18: 99
5) 16: 36	6) 64: 72
7) 4: 60	8) 70: 40
9) 8: 64	10) 16: 24

Score: ....

Answer Key	
1) 7: 2	2) 2: 5
3) 7: 5	4) 2: 11
5) 4: 9	6) 8: 9
7) 1: 15	8) 7: 4
9) 1: 8	10) 2: 3

Name: ....

Create a Proportion

- ✓ A proportion carries two equal fractions! A proportion means equality of two fractions.
- ✓ If you want to create a proportion, simply find (or create) two equal fractions.

EXAMPLE:

Explain if these ratios form a proportion. $\frac{3}{5}$ and $\frac{24}{45}$

   Use cross multiplication: $\frac{3}{5} = \frac{24}{45} \rightarrow 3 \times 45 = 5 \times 24 \rightarrow 135 = 120$, which is not correct.

   Thus, this pair of ratios doesn't form a proportion.

PRACTICES:

Create proportion from the given set of numbers.

1) 3, 2, 9, 6	2) 4, 18, 12, 6
3) 5, 11, 25, 55	4) 24, 7, 21, 8
5) 49, 7, 12, 84	6) 15, 12, 30, 24
7) 20, 10, 200, 1	8) 9, 27, 81, 3
9) 4, 2, 16, 32	10) 9, 6, 27, 18

Score: ....

Answer Key	
1) 2: 6 = 3: 9	2) 4: 12 =6: 18
3) 5: 25 = 11: 55	4) 8: 24 =7: 21
5) 7: 49 = 12: 84	6) 12: 24 =15: 30
7) 1: 10 = 20: 200	8) 3: 27 =9: 81
9) 2: 16=4: 32	10) 6: 18 = 9: 27

Name: ....

Similar Figures

✓ Two or more figures are equivalent if their corresponding angles are equal, and the corresponding sides are in proportion.

EXAMPLE:

4–5–6 triangle is like an 8–10–12 triangle.

PRACTICES:

Each pair of figures is similar. Find the missing side.

1)

2)

3)

4)

5)

    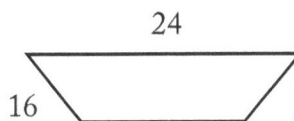

6)

7)

8)

9)

10)

Score: ....

Answer Key	
1) 6	2) 4
3) 3	4) 5
5) 2	6) 3
7) 1	8) 8
9) 20	10) 18

Name: ....

Ratio and Rates Word Problems

✓ To solve a rate word problem or a ratio, create a proportion and then use cross multiplication method.

EXAMPLE:

A tree **32 feet** tall has a shadow **12 feet** long. Jack is **6 feet** tall. How long is Jack's shadow?

To solve for the missing number, write in a proportion.

$\frac{32}{12} = \frac{6}{x} \rightarrow 32x = 6 \times 12 = 72$

$32x = 72 \rightarrow x = \frac{72}{32} = 2.25$

PRACTICES:

Solve.

1) In a party, 8 soft drinks are required for every 35 guests. If there are 560 guests, how many soft drinks is required?

2) You can buy 6 cans of green beans at a supermarket for $3.50. How much does it cost to buy 42 cans of green beans?

3) The price of 5 bananas at the first Market is $1.05. The price of 7 of the same bananas at second Market is $1.07. Which place is the better buy?

4) In Peter's class, 21 of the students are tall and 9 are short. In Elise's class 56 students are tall and 24 students are short. Which class has a higher ratio of tall to short students?

5) The bakers at a Bakery can make 110 bagels in 4 hours. How many bagels can they bake in 6 hours? What is that rate per hour?

6) A certain sweet recipe calls for 3 kg of sugar for every 6 kg of flour. If 63 kg of this sweet must be prepared, how much sugar is required?

7) In a mixture of 45 liters, the ratio of sugar solution to salt solution is 1:2. What is the amount of sugar solution to be added if the ratio must be 2:1?

8) In a bag of red and green sweets, the ratio of red sweets to green sweets is 3:4. If the bag contains 120 green sweets, how many red sweets are there?

9) If the ratio of chocolates to ice-cream cones in a box is 5:8 and the number of chocolates is 30, find the number of ice-cream cones.

10) In a group, the ratio of doctors to lawyers is 5:4. If the total number of people in the group is 72, what is the number of lawyers in the group?

Score: ....

Answer Key

1) 128	2) $24.5
3) The price at the second Market is a better buy.	4) The ratio for both classes equal 7 to 3.
5) 165, the rate is 27.5 per hour.	6) 21 kg (3+6=9, $\frac{63}{9} = 7$. Therefore, 3:6=21:42)
7) 45	8) 90
9) 48	10) 32

Name: ....

Percentage Calculations

- ✓ Percent is called the ratio of a number and 100. It always possesses the same denominator, 100. The symbol used for percent is %.
- ✓ Percent is another method to write decimals or fractions. For example:
$$40\% = 0.40 = \frac{40}{100} = \frac{2}{5}$$
- ✓ Use the given formula to find part, whole, or percent:
$$\text{part} = \frac{\text{percent}}{100} \times \text{whole}$$

EXAMPLE:

What is **10%** of **45**?

Use this formula: $\text{part} = \frac{\text{percent}}{100} \times \text{whole}$

$\text{part} = \frac{10}{100} \times 45 \rightarrow \text{part} = \frac{1}{10} \times 45 \rightarrow \text{part} = \frac{45}{10} \rightarrow \text{part} = 4.5$

PRACTICES:

Calculate the percentages.

1) 75% of 45	2) 50% of 66
3) 90% of 58	4) 25% of 88
5) 5% of 100	6) 80% of 60

Solve.

7) What percentage of 60 is 6	8) 6.76 is what percentage of 52?
9) 17 is what percentage of 85?	10) Find what percentage of 96 is 24.

Score: ....

Answer Key	
1) 33.75	2) 33
3) 52.2	4) 22
5) 5	6) 48
7) 10%	8) 13%
9) 20%	10) 25%

Name: ....

Percent Problems

- ✓ In each percent question, we are finding the base, or part or the percent.
- ✓ Use the following equations to find each missing portion.
  - Base = Part ÷ Percent
  - Part = Percent × Base
  - Percent = Part ÷ Base

EXAMPLE:

20 is 5% of what number?

Use the formula: $Base = Part \div Percent \rightarrow Base = 20 \div 0.05 = 400$

20 is 5% of 400

PRACTICES:

Solve each problem.

1) 52% of what number is 13?	2) What is 15% of 9 inches?
3) What percent of 185.6 is 23.2?	4) 24 is 72% of what?
5) 35 is what percent of 70?	6) 10 is 200% of what?
7) 14 is what percent of 70?	8) 26% of 100 is what number?

9) Mia requires 50% to pass. If she gets 250 marks and falls short by 90 marks, what were the maximum marks she could have got?

10) Jack scored 14 out of 70 marks in mathematics, 9 out of 10 marks in history and 56 out of 100 marks in science. In which subject his percentage of marks is the best?

Score: ....

Answer Key	
1) 25	2) 1.35
3) 12.5	4) 33.33
5) 50%	6) 5
7) 20%	8) 26
9) 680	10) History

Name: ....

Markup, Discount, and Tax

- ✓ Markup = selling price – cost
- ✓ Markup rate = markup is divided by the cost
- ✓ Discount = Multiply the rate of discount by regular price.
- ✓ Tax: To find tax, multiply the taxable amount (income, property value, etc.) to the tax rate.
- ✓ To find tip, multiply selling price to the rate.

EXAMPLE:

With an **10**% discount, Ella was able to save **$20** on a dress. What was the original price of the dress?

$10\% \ of \ x = 20, \frac{10}{100} \times x = 20, x = \frac{100 \times 20}{10} = 200$

PRACTICES:

Find the selling price of each item.

1) Cost of a chair: $20, markup: 30%, discount: 10%, tax: 10%

2) Cost of computer: $1,600.00, markup: 65%

3) Cost of a pen: $3.20, markup: 50%, discount: 15%, tax: 5%

4) Cost of a puppy: $1,800, markup: 40%, discount: 10%

5) Cost of a book: $50, markup: 40%, discount: 20%, tax: 5%

6) Original price of a tablet: $400, discount: 20% Tax: 5%,

7) Original price of a book: $50, markup:20% Discount: 20%, Tax: 2.5%,

8) Original price of a cellphone: $500, markup:14% Discount: 25%, Tax: 1.6%,

9) Original price of a sofa: $800, markup:10% Discount: 15%, Tax: 1.5%,

10) Original price of a car: $40,000, markup:12% Discount: 25%, Tax: 6.5%,

Score: ....

Answer Key	
1) $25.74	2) $2,640
3) $4.284	4) $2,268
5) $58.8	6) $336
7) $49.2	8) $434.34
9) $759.22	10) $35,784

Name: ....

Simple Interest

- ✓ Simple Interest: The charge for borrowing money or the return for lending it.

  To solve a simple interest problem, use this formula:

- ✓ Interest = principal x rate x time $\Rightarrow I = p \times r \times t$

EXAMPLE:

Find simple interest for $5,200 at 4% for 3 years.

Use Interest formula: $I = prt$

$P = \$5,200$, r $= 4\% = \dfrac{4}{100} = 0.04$ and $t = 3$

Then: $I = 5,200 \times 0.04 \times 3 = \$624$

PRACTICES:

Use simple interest to find the ending balance.

1) $1,200 at 15% for 3 years.	2) $320,000 at 2.85% for 7 years.
3) $1,500 at 2.25% for 12 years.	4) $12,500 at 6.2% for 4 years.
5) $31,000 at 1.5% for 10 months.	6) $18,000 at 5.2% for 5 years.

7) Emily puts $6,000 into an investment yielding 3.25% annual simple interest; she left the money in for 3 years. How much interest does Sara get at the end of those 3 years?

8) A new car, valued at $42,000, depreciates at 7.5% per year from original price. Find the value of the car 6 years after purchase.

9) $880 interest is earned on a principal of $2,200 at a simple interest rate of 4% interest per year. For how many years was the principal invested?

10) A bank is offering 3.2% simple interest on a savings account. If you deposit $15,000, how much interest will you earn in six years?

Score: ....

Answer Key

1) $1,740	2) $383,840.00
3) $1,905.00	4) $15,600
5) $31,387.50	6) $22,680
7) $585.00	8) $23,100
9) 10 years	10) $2,880

Name: ....

Converting Between Percent, Fractions, and Decimals

✓ To a percent: We move the decimal point 2 places to the right and add the percentage (%) symbol.

✓ Divide by 100 to change a number from percent to decimal.

EXAMPLE:

$30\% = 0.30$

$0.24 = 24\%$

PRACTICES:

Converting fractions to decimals.	Write each decimal as a percent.
1) $\frac{23}{10}$	2) 0.002
3) $\frac{2}{20}$	4) 0.08
5) $\frac{7}{100}$	6) 0.2
7) $\frac{20}{50}$	8) 3.25
9) $\frac{3}{60}$	10) 1.01

Score: ....

Answer Key

1) 2.3	2) 0.2%
3) 0.1	4) 8%
5) 0.07	6) 20%
7) 0.4	8) 325%
9) 0.05	10) 101%

Chapter 4 : Exponents and Radicals

Topics that you'll learn in this chapter:

- ➤ Multiplication Property of Exponents

- ➤ Division Property of Exponents

- ➤ Powers of Products and Quotients

- ➤ Zero, Negative Exponents and Bases

"Mathematics is no more computation than typing is literature." – *John Allen Paulos*

Name: ....

Multiplication Property of Exponents

- ✓ Exponents are shorthand for recurrent multiplication of the identical number by itself. For example, instead of writing 2×2, we can write 2^2. For $3 \times 3 \times 3 \times 3$, we can write 3^4

- ✓ In algebra, a variable is a letter used as a replacement for a number. The most common letters are: $x, y, z, a, b, c, m,$ and n.

- ✓ Exponent's rules: $(x^a)^b = x^{a \times b}$, $(xy)^a = x^a \times y^a$,

  $x^a \times x^b = x^{a+b}$, $x^a \times y^a = (xy)^a$,

EXAMPLE:

Multiply. $-2x^5 \times 7x^3 =$

Use Exponent's rules: $x^a \times x^b = x^{a+b} \rightarrow x^5 \times x^3 = x^{5+3} = x^8$

Then: $-2x^5 \times 7x^3 = -14x^8$

PRACTICES:

Simplify.

1) $4^3 \times 4^2$	2) $2 \times 2^2 \times 2^3$
3) $2^4 \times 2$	4) $8x^2 \times x$
5) $15x^7 \times x$	6) $3x \times x^3$
7) $2x^5 \times 5x^4$	8) $5x^2 \times 3x^2y^2$
9) $6y^5 \times 8xy^2$	10) $5xy^3 \times 4x^3y^2$

Score: ....

Answer Key	
1) 4^5	2) 2^6
3) 2^5	4) $8x^3$
5) $15x^8$	6) $3x^4$
7) $10x^9$	8) $15x^4y^2$
9) $48xy^7$	10) $20x^4y^5$

Name: ....

Division Property of Exponents

✓ For division of exponents, we can use these formulas: $\left(\frac{a}{b}\right)^c = \frac{a^c}{b^c}$, $b \neq 0$

$$\frac{x^a}{x^b} = x^{a-b}, x \neq 0 \qquad \frac{x^a}{y^a} = \left(\frac{x}{y}\right)^a, y \neq 0$$

$$\frac{x^a}{x^b} = \frac{1}{x^{b-a}}, x \neq 0, \qquad \frac{1}{x^b} = x^{-b}$$

EXAMPLE:

Simplify. $\frac{4x^3y}{36x^2y^3} =$

First you cancel the common factor: $4 \rightarrow \frac{4x^3y}{36x^2y^3} = \frac{x^3y}{9x^2y^3}$

Use Exponent's rules: $\frac{x^a}{x^b} = x^{a-b} \rightarrow \frac{x^3}{x^2} = x^{3-2}$

Then: $\frac{4x^3y}{36x^2y^3} = \frac{xy}{9y^3} \rightarrow$ now cancel the common factor: $y \rightarrow \frac{xy}{9y^3} = \frac{x}{9y^2}$

PRACTICES:

Simplify.

1) $\frac{4^3}{4}$	2) $\frac{51}{51^{14}}$
3) $\frac{5^2}{5^3}$	4) $\frac{3^4}{15^4}$
5) $\frac{x}{x^7}$	6) $\frac{42x^2}{6x^2}$
7) $\frac{3x^{-3}}{12x^{-1}}$	8) $\frac{81x^5}{9x^3}$
9) $\frac{3x^4}{4x^5}$	10) $\frac{21x}{3x^2}$

Score: ....

Answer Key	
1) 4^2	2) $\dfrac{1}{51^{13}}$
3) $\dfrac{1}{5}$	4) $\left(\dfrac{1}{5}\right)^4 = \dfrac{1}{5^4}$
5) $\dfrac{1}{x^6}$	6) 7
7) $\dfrac{1}{4x^2}$	8) $9x^2$
9) $\dfrac{3}{4x}$	10) $\dfrac{7}{x}$

Name: ....

Powers of Products and Quotients

✓ For any number except zero, a and b and any integer x, $(ab)^x = a^x \times b^x$.

EXAMPLE:

Simplify. $(3x^5y^4)^2 =$

Use Exponent's rules: $(x^a)^b = x^{a \times b}$

$$(3x^5y^4)^2 = (3)^2(x^5)^2(y^4)^2 = 9x^{5 \times 2}y^{4 \times 2} = 9x^{10}y^8$$

PRACTICES:

Simplify.

1) $(5x^3)^2$

2) $(xy)^2$

3) $(ax^2)^3$

4) $(2x^3yz)^2$

5) $(4x^2y^3)^2$

6) $(5x^2y^3)^2$

7) $(2xy^2)^3$

8) $(2x^3y)^4$

9) $(7x^4y^8)^2$

10) $(10x)^3$

Score: ....

Answer Key	
1) $25x^6$	2) x^2y^2
3) a^3x^6	4) $4x^6y^2z^2$
5) $16x^4y^6$	6) $25x^4y^6$
7) $8x^3y^6$	8) $8x^{12}y^4$
9) $49x^8y^{16}$	10) $1,000x^3$

Name: ....

Zero and Negative Exponents

✓ A negative exponent just means that the base is on the wrong side of the fraction line, so you need to flip the base to the other side. For example, "x^{-2}" (pronounced as "ecks to the minus two") just means "x^2" but below, as in $\frac{1}{x^2}$

EXAMPLE:

Evaluate. $\left(\frac{4}{9}\right)^{-2} =$

Use Exponent's rules: $\frac{1}{x^b} = x^{-b} \rightarrow \left(\frac{4}{9}\right)^{-2} = \frac{1}{\left(\frac{4}{9}\right)^2} = \frac{1}{\frac{4^2}{9^2}}$

Now use fraction rule: $\frac{1}{\frac{b}{c}} = \frac{c}{b} \rightarrow \frac{1}{\frac{4^2}{9^2}} = \frac{9^2}{4^2} = \frac{81}{16}$

PRACTICES:

Evaluate the following expressions.

1) 4^{-2}	2) 5^{-2}
3) 6^{-2}	4) 3^{-4}
5) 10^{-1}	6) 33^{-1}
7) 6^{-1}	8) 3^{-2}
9) 9^{-2}	10) 4^{-1}

Score: ....

Answer Key	
1) $\frac{1}{16}$	2) $\frac{1}{25}$
3) $\frac{1}{36}$	4) $\frac{1}{81}$
5) $\frac{1}{10}$	6) $\frac{1}{33}$
7) $\frac{1}{6}$	8) $\frac{1}{9}$
9) $\frac{1}{81}$	10) $\frac{1}{4}$

Name: ....

Negative Exponents and Negative Bases

✓ First make the power positive. A negative exponent can be written as reciprocal of that number with a positive exponent.

✓ The parenthesis is significant!

✓ 5^{-3} is not the same as $(-5)^{-3}$

$(-5)^{-3} = -\frac{1}{5^3}$ and $(5)^{-3} = +\frac{1}{5^3}$

EXAMPLE:

Simplify. $(-\frac{5x}{3yz})^{-3} =$

Use Exponent's rules: $\frac{1}{x^b} = x^{-b} \rightarrow (-\frac{5x}{3yz})^{-3} = \frac{1}{(-\frac{5x}{3yz})^3} = \frac{1}{-\frac{5^3x^3}{3^3y^3z^3}}$

Now use fraction rule: $\frac{1}{\frac{b}{c}} = \frac{c}{b} \rightarrow \frac{1}{-\frac{5^3x^3}{3^3y^3z^3}} = -\frac{3^3y^3z^3}{5^3x^3} = -\frac{27y^3z^3}{125x^3}$

PRACTICES:

Simplify.

1) 7^{-1}	2) $-2x^{-2}$
3) $\frac{x}{x^{-5}}$	4) $-\frac{a^{-2}}{b^{-1}}$
5) $\frac{7}{x^{-5}}$	6) $\frac{2b}{-5c^{-2}}$
7) $\frac{2n^{-1}}{12p^{-2}}$	8) $\frac{8b^{-4}}{3c^{-2}}$
9) $89xy^{-2}$	10) $(\frac{1}{3})^{-2}$

Score: ....

Answer Key	
1) $\frac{1}{7}$	2) $-\frac{2}{x^2}$
3) x^5	4) $-\frac{b^1}{a^2}$
5) $7x^5$	6) $-2\frac{bc^2}{5}$
7) $\frac{p^2}{6n}$	8) $\frac{8c^2}{3b^4}$
9) $\frac{89x}{y^2}$	10) 9

Name: ....

Writing Scientific Notation

✓ It is used to write very big or very small values in decimal representation.

✓ In scientific notation all numbers can be written in the form of:

$m \times 10^n$, where $1 \leq m \leq 10$ and n is any integer.

Decimal notation	Scientific notation
5	5×10^0
$-25,000$	-2.5×10^4
0.5	5×10^{-1}
2,122.456	2.122456×10^3

EXAMPLE:

Write 8.3×10^{-5} in standard notation.

$10^{-5} \rightarrow$ When the decimal moved to the right, the exponent is negative.

Then: $8.3 \times 10^{-5} = 0.000083$

PRACTICES:

Write each number in scientific notation.

1) 12,000,000	2) 25×10^5
3) 0.0015	4) 54,000
5) 0.0005021	6) 666,012
7) 0.00000076	8) 102,900,000
9) 4,100,000,000	10) 3,600,000

Score: ....

Answer Key

1) 1.2×10^7	2) 2.5×10^6
3) 1.5×10^{-3}	4) 5.4×10^4
5) 5.021×10^{-4}	6) 6.66012×10^5
7) 7.6×10^{-7}	8) 1.029×10^8
9) 4.1×10^9	10) 3.6×10^6

Name: ....

Square Roots

✓ A square root of x is a number p whose square is: $p^2 = x$

p is a square root of x.

EXAMPLE:

Find the square root of $\sqrt{225}$.

First factor the number: $225 = 15^2$, Then: $\sqrt{225} = \sqrt{15^2}$

Now use radical rule: $\sqrt[n]{a^n} = a$

Then: $\sqrt{15^2} = 15$

PRACTICES:

Find the value each square root.

1) $\sqrt{25}$	2) $\sqrt{1,600}$
3) $\sqrt{100}$	4) $\sqrt{121}$
5) $\sqrt{4}$	6) $\sqrt{225}$
7) $\sqrt{10,000}$	8) $\sqrt{16}$
9) $\sqrt{64}$	10) $\sqrt{36}$

Score: ....

Answer Key	
1) 5	2) 40
3) 10	4) 11
5) 2	6) 15
7) 100	8) 4
9) 8	10) 6

Chapter 5 : Algebraic Expressions

Topics that you'll learn in this chapter:

- ➤ Expressions and Variables

- ➤ Simplifying Variable and Polynomial Expressions

- ➤ Translate Phrases into an Algebraic Statement

- ➤ The Distributive Property

- ➤ Evaluating One and two Variable

- ➤ Combining like Terms

"Without mathematics, there's nothing you can do. Everything around you are mathematics. Everything around you are numbers." – Shakuntala Devi

Name: ....

Translate Phrases into an Algebraic Statement

How to translate key words and phrases into algebraic expressions:

- ✓ Addition: the sum of, more than, plus, etc.
- ✓ Subtraction: less than, decreased, minus, etc.
- ✓ Multiplication: times, multiplied, product, etc.
- ✓ Division: quotient, ratio, divided, etc.

EXAMPLE:

5 times the sum of x and 8

Sum of 8 and x: $8 + x$. Times is used for multiplication. Then: $5 \times (8 + x)$

PRACTICES:

Write an algebraic expression for each phrase.

1) Fifteen subtracted from a number.

2) The quotient of seventeen and a number.

3) A number increased by fifty.

4) A number divided by -21.

5) The difference between sixty –three and a number.

6) Threefold a number decreased by 45.

7) Seven times the sum of a number and -21.

8) The quotient of 90 and the product of a number and -8.

9) Nine subtracted from 4 times a number.

10) The difference of six and a number.

Score: ....

Answer Key	
1) $x - 15$	2) $\dfrac{17}{x}$
3) $x + 50$	4) $-\dfrac{x}{21}$
5) $63 - x$	6) $3x - 45$
7) $7(x + (-21))$	8) $-\dfrac{90}{8x}$
9) $4x - 9$	10) $6 - x$

Name: ....

The Distributive Property

✓ Distributive Property:

$$a(b + c) = ab + ac$$

EXAMPLE:

Simply. $(5x - 3)(-5) =$

Use Distributive Property formula: $a(b + c) = ab + ac$

$$(5x - 3)(-5) = -25x + 15$$

PRACTICES:

Use the distributive property to simply each expression.

1) $4(9 - 3x)$	2) $-(-8 - 4x)$
3) $(-5x - 1)(-2)$	4) $(-3)(2x - 4)$
5) $4(5 + 3x)$	6) $(-9x + 10)3$
7) $(-4 - 5x)(-3)$	8) $(-2x)(-3 + 2x) - 3x(1 - 5x)$
9) $(-2)(3x - 1) + 4(3x + 2)$	10) $(-15)(2x + 3)$

Score: ....

Answer Key

1) $-12x + 36$	2) $4x + 8$
3) $10x + 2$	4) $-6x + 12$
5) $12x + 20$	6) $-27x + 30$
7) $15x + 12$	8) $11x^2 + 3x$
9) $6x + 10$	10) $-30x - 45$

Name: ....

Evaluating One Variable

- ✓ To solve a variable expression, find the variable and substitute a number for that variable.
- ✓ Perform the mathematical operations.

EXAMPLE:

Solve this expression. $12 - 2x , x = -1$

First substitute -1 for x, then:

$$12 - 2x = 12 - 2(-1) = 12 + 2 = 14$$

PRACTICES:

Simplify each algebraic expression.

1) $5x + 4, x = 1$	2) $x + (-4), x = -6$
3) $-10x + 8, x = -2$	4) $\left(-\frac{36}{x}\right) - 10 + 2x, x = 6$
5) $\frac{36}{x} - 3, x = 3$	6) $(-10) - \frac{x}{4} + 4x, x = -8$
7) $15 + 6x - 3, x = -1$	8) $(-5) + \frac{x}{8}, x = 64$
9) $\left(-\frac{24}{x}\right) - 10 + 5x, x = 4$	10) $(-4) + \frac{4x}{9}, x = 81$

Score: ....

Answer Key	
1) 9	2) −10
3) 28	4) −4
5) 9	6) −40
7) 6	8) 3
9) 4	10) 32

Name: ....

Evaluating Two Variables

To solve an algebraic expression, substitute a number for each variable and perform the mathematical operations.

EXAMPLE:

Solve this expression. $-3x + 5y$, $x = 2$, $y = -1$

First substitute 2 for x, and -1 for y, then:

$$-3x + 5y = -3(2) + 5(-1) = -6 - 5 = -11$$

PRACTICES:

Simplify each algebraic expression.

1) $5a - (5 - b)$, $a = 2, b = 3$	2) $5x + 3y - 6 + 3y$, $x = 3, y = 1$
3) $\left(-\frac{27}{x}\right) + 4 + 3y$, $x = 3, y = 5$	4) $(-4)(-3a - 5b)$, $a = 3, b = 4$
5) $7x + 10 - 5y$, $x = 3, y = 6$	6) $18 + 3(-x - 4y)$, $x = 2, y = 5$
7) $12x + 2y$, $x = 5, y = 10$	8) $x \times 6 \div 3y$, $x = 6, y = 1$
9) $4x - 3y$, $x = 6, y = 3$	10) $\left(-\frac{14}{x}\right) + 4y$, $x = 7, y = -3$

Score: ....

Answer Key	
1) 8	2) 15
3) 10	4) 116
5) 1	6) −48
7) 80	8) 12
9) 15	10) −14

Name: ....

Expressions and Variables

- ✓ In algebra, a variable is a letter used as a replacement for a number. The most common letters are: $x, y, z, a, b, c, m, and\ n$.
- ✓ An algebraic expression is an expression that has variables, integers, and math operations such as addition, subtraction, multiplication, division, etc.
- ✓ In an expression, we can combine "identical" terms. (Values with same power and variable)

EXAMPLE:

Simplify this expression. $(10x + 2x + 3) =?$

Combine like terms. Then: $(10x + 2x + 3) = 12x + 3$ (remember you cannot combine variables and numbers

PRACTICES:

Simplify each expression.

1) $10(-3 - 8x), x = 4$	2) $-3(5 - 8x) - 6x, x = 1$
3) $2x - 8x, x = 2$	4) $x + 12x, x = 6$
5) $20 - 5x + 10x + 5, x = 3$	6) $15(5x + 3), x = 0$
7) $20(4 - x) - 9, x = 2$	8) $20x - 8x - 10, x = 5$
9) $6x + 9y,\ x = 4, y = 2$	10) $6x - 2x, x = 8$

Score: ....

Answer Key	
1) − 350	2) 3
3) − 12	4) 78
5) 40	6) 45
7) 31	8) 50
9) 42	10) 32

Name: ....

Combining like Terms

✓ We separate the terms by "+" and "−" signs.

✓ Identical terms are those terms with same powers and same variables.

Make sure to use the "+" or "−" that is in front of the coefficient

EXAMPLE:

Simplify this expression. $(-5)(8x - 6) =$

Use Distributive Property formula: $a(b + c) = a + ac$

$(-5)(8x - 6) = -40x + 30$

PRACTICES:

Simplify each expression.

1) $-8(-5x + 1)$	2) $6(-2 + 4x)$
3) $-8 - 14x + 16x + 3$	4) $9x - 7x - 15 + 18$
5) $(-9)(12x - 21) + 31$	6) $2(4x + 9) + 12x$
7) $4(-2x - 17) + 14(3x + 1)$	8) $(9x - 5y)7 + 25y$
9) $4.5x^3 \times (-8x)$	10) $-19 - 15x^2 + 12x^2$

Score: ....

Answer Key

1) $40x - 8$	2) $24x - 12$
3) $2x - 5$	4) $2x + 3$
5) $220 - 108x$	6) $20x + 18$
7) $34x - 54$	8) $63x - 10y$
9) $-36x^4$	10) $-3x^2 - 19$

Name: ....

Simplifying Polynomial Expressions

✓ A polynomial is a unique expression that consists of coefficients and variables that performs only the arithmetic of addition, subtraction, multiplication, and non-negative integer exponents of variables.

$$P(x) = a_n x^n + a_{n-1} x^{n-1} + \ldots + a_2 x^2 + a_1 x + a_0$$

EXAMPLE:

Simplify this Polynomial Expressions. $4x^2 - 5x^3 + 15x^4 - 12x^3 =$

Combine "like" terms: $-5x^3 - 12x^3 = -17x^3$

Then: $4x^2 - 5x^3 + 15x^4 - 12x^3 = 4x^2 - 17x^3 + 15x^4$

Then write in standard form: $4x^2 - 17x^3 + 15x^4 = 15x^4 - 17x^3 + 4x^2$

PRACTICES:

Simplify each polynomial.

1) $(2x^2 + 4) - (9 + 5x^2)$	2) $(25x^3 - 12x^2) - (6x^2 - 9x^3)$
3) $14x^5 - 15x^6 + 2x^5 - 16x^6 + x^6$	4) $(15 + 12x^3) + (3x^3 + 5)$
5) $13x^3 - 15x^4 + 12x^3 + 20x^4$	6) $-6x^2 + 15x^2 + 17x^3 + 16 - 32$
7) $15x^3 + 12 + 2x^2 - 5x - 10x$	8) $24x^2 - 16x^3 - 4x(2x^2 + 3x)$
9) $(21x^4 - 10x) - (2x - x^4)$	10) $(7x^2 - 9) + (x^2 - 8x^3)$

Score: ....

Answer Key	
1) $-3x^2 - 5$	2) $34x^3 - 18x^2$
3) $-30x^6 + 16x^5$	4) $15x^3 + 20$
5) $5x^4 + 25x^3$	6) $17x^3 + 9x^2 - 16$
7) $15x^3 + 2x^2 - 15x + 12$	8) $-24x^3 + 12x^2$
9) $22x^4 - 12x$	10) $-8x^3 + 8x^2 - 9$

Chapter 6 : Equations and Inequalities

Topics that you'll learn in this chapter:

- One, Two, and Multi – Step Equations

- Graphing Single– Variable Inequalities

- One, Two, and Multi – Step Inequalities

- Solving Systems of Equations by Substitution and Elimination

- Finding Slope and Writing Linear Equations

- Graphing Lines Using Slope– Intercept and Standard Form

- Graphing Linear Inequalities

- Finding Midpoint and Distance of Two Points

"The study of mathematics, like the Nile, begins in minuteness but ends in magnificence." –
Charles Caleb Colton

Name: ....

One–Step Equations

✓ The values of two algebraic expressions on both sides of an equation are always equal.

$$ax + b = c$$

✓ You only require performing one Math operation to solve the problem.

EXAMPLE:

Solve this equation. $x + 24 = 0$, $x = ?$

Here, we have the addition operation, and its inverse operation is subtraction. To solve this equation, subtract 24 from both sides of the equation: $x + 24 - 24 = 0 - 24$

Then simplify: $x + 24 - 24 = 0 - 24 \rightarrow x = -24$

PRACTICES:

Solve each equation.

1) $x + 4 = 16$	2) $48 = (-2) + x$
3) $5x = (-105)$	4) $(-8) = (8x)$
5) $(-2) = 14 + x$	6) $5 + x = 6$
7) $2x + 3 = (-7)$	8) $28 = x + 7$
9) $(-15) + x = (-15)$	10) $12x = (-36)$

Score: ....

Answer Key	
1) 12	2) 50
3) −21	4) −1
5) −16	6) 1
7) −5	8) 21
9) 0	10) −3

Name: ....

Two–Step Equations

- ✓ You only require performing two math operations (add, subtract, multiply, or divide) to solve the equation.
- ✓ Simplify the equation using the inverse of addition or subtraction.
- ✓ Simplify the equation further by using the inverse of division or multiplication.

EXAMPLE:

Solve this equation. $3x = 15, x =?$

Here, we have the multiplication operation (variable x is multiplied by 3) and its inverse operation is division. To solve this problem, we divide both sides of equation by 3:

$3x = 15 \rightarrow 3x \div 3 = 15 \div 3 \rightarrow x = 5$

PRACTICES:

Solve each equation.

1) $4(2 + 2x) = 8$	2) $(-5)(x - 3) = 25$
3) $(-5)(2x - 5) = (-15)$	4) $4(9 + 3x) = -12$
5) $6(2x + 1) = 30$	6) $2(x + 2) = 42$
7) $2(12 + 6x) = 60$	8) $(-10)(5x) = 100$
9) $4(3x + 3) = 24$	10) $\dfrac{x - 5}{3} = 4$

Score: ....

Answer Key	
1) 0	2) –2
3) 4	4) –4
5) 2	6) 10
7) 3	8) –2
9) 1	10) 17

Name: ....

Multi–Step Equations

✓ Combine "identical" terms on one side.

✓ Put variables to one side by adding or subtracting.

✓ Simplify the equation by using the inverse of addition or subtraction.

✓ Simplify further by using the inverse of division or multiplication.

EXAMPLE:

Solve this equation. $-(2 - x) = 5$

First, use Distributive Property: $-(2 - x) = -2 + x$

Now by adding 2 to both sides of the equation, we can solve it.

$-2 + x = 5 \rightarrow -2 + x + 2 = 5 + 2$
Now simplify: $-2 + x + 2 = 5 + 2 \rightarrow x = 7$

PRACTICES:

Solve each equation.

1) $8 - 2x = 28$	2) $-10 = -(x + 7)$
3) $2x - 17 = (-x) + 1$	4) $-2x = (-3x) - 8$
5) $5(14 + 2x) + 3x = -x$	6) $x - 11 = x - 5 + 2x$
7) $15 + 2x = (-25) - 2x + 3x$	8) $-3(x - 3x) = 40 - 4x$
9) $24 + 8x + x = (-x + 4)$	10) $-8(1 + 5x) = 152$

Score: ....

Answer Key	
1) -10	2) 3
3) 6	4) -8
5) -5	6) -3
7) -40	8) 4
9) -2	10) -4

Name: ....

Graphing Single–Variable Inequalities

- ✓ Inequality is like equations and uses symbols for "less than" (<) and "greater than" (>).
- ✓ To solve inequalities, we need to separate the variable. (Like in equations)
- ✓ Find the value of the inequality on the number line to graph an inequality.
- ✓ For greater than or less than draw open circle on the value of the variable.
- ✓ Use filled circle if there is an equal sign too.
- ✓ Draw a line to the left or to the right for less or greater than.

EXAMPLE:

Draw a graph for $x > 2$

Since, the variable is greater than 2, then we need to find 2 and draw an open circle above it. Then, draw a line to the right.

PRACTICES:

Draw a graph for each inequality.

1) $2 \geq x$

2) $x < 3$

3) $5 \geq x$

4) $x \geq -2$

5) $x > 0$

6) $-1.5 < x$	
7) $x \geq -1$	

Score: ....

Answer Key

1) $2 \geq x$	
2) $x < 3$	
3) $5 \geq x$	
4) $x \geq -2$	
5) $x > 0$	
6) $-1.5 < x$	
7) $x \geq -1$	

Name: ....

One–Step Inequalities

- ✓ Like equations, first separate the variable by using inverse operation.
- ✓ For dividing or multiplying both sides by negative numbers, flip the direction of the inequality sign.

EXAMPLE:

Solve this inequality. $x - 1 \leq 2$

Add 1 to both sides. $x - 1 \leq 2 \rightarrow x - 1 + 1 \leq 2 + 1$, then: $x \leq 3$

PRACTICES:

Solve each inequality and graph it.

1) $2x + 3 \geq 7$

2) $x < 3$

3) $5 \geq x$

4) $x \geq -2$

5) $x > 0$

6) $-1.5 < x$

7) $x \geq -1$

Score: ....

Answer Key

1) $2 \geq x$

2) $x < 3$

3) $5 \geq x$

4) $x \geq -2$

5) $x > 0$

6) $-1.5 < x$

7) $x \geq -1$

Name: ....

Two–Step Inequalities

- ✓ Separate the variable.
- ✓ Flip the direction of the inequality sign for dividing both sides by negatives numbers.
- ✓ We can simplify by using the inverse of addition or subtraction.
- ✓ We can simplify further by using the inverse of division or multiplication.

EXAMPLE:

Solve: $2x + 9 \geq 11$

First add -9 to both sides: $2x + 9 - 9 \geq 11 - 9 \to 2x \geq 2$

Now, divide both sides by 2: $2x \geq 2 \to x \geq 1$

PRACTICES:

Solve each inequality and graph it.

1) $x - 4 \leq 4$	2) $x + 4 \geq 5$
3) $3x - 2 \leq 7$	4) $5x + 2 < 12$
5) $x + 7 \geq 9$	6) $3x - 3 \leq 3$
7) $7x - 4 < 3$	8) $8 + x \leq 13$
9) $2x + 7 \leq 11$	10) $10x - 16 < 4$

Score: ....

Answer Key

1) $x \leq 8$	2) $x \geq 1$
3) $x \leq 3$	4) $x < 2$
5) $x \geq 2$	6) $x \leq 2$
7) $x < 1$	8) $x \leq 5$
9) $x \leq 2$	10) $x < 2$

Name: ....

Multi–Step Inequalities

- ✓ Separate the variable.

- ✓ We can simplify by using the inverse of addition or subtraction.

- ✓ We can simplify further by using the inverse of division or multiplication.

EXAMPLE:

Solve this inequality. $2x - 2 \leq 6$

First add 2 to both sides: $2x - 2 + 2 \leq 6 + 2 \rightarrow 2x \leq 8$

Now, divide both sides by 2: $2x \leq 8 \rightarrow x \leq 4$

PRACTICES:

Solve each inequality.

1) $-(x + 3) + 8 < 25$	2) $\frac{3x + 1}{2} \leq 5$
3) $\frac{x - 4}{3} > 7$	4) $4(x - 2) \leq 8$
5) $\frac{x}{3} + \frac{1}{3} < 2$	6) $\frac{x+4}{5} > 3$
7) $\frac{x}{8} + \frac{3}{4} < 1$	8) $2(x + 5) + 4 > 10$
9) $24 + 5x < 4$	10) $\frac{x+2}{3} > 10$

Score: ....

Answer Key	
1) $x > -20$	2) $x \leq 3$
3) $x > 25$	4) $x \leq 4$
5) $x < 5$	6) $x > 11$
7) $x < 2$	8) $x > -2$
9) $x < -4$	10) $x > 28$

Name: ....

Solving Systems of Equations by Substitution

✓ Let the system of equations. $x + y = 1; -2x + y = 4$

Put $x = 1 - y$ in the second equation.

$-2(1 - y) + y = 4 \rightarrow -2 + 2y + y = 4 \Rightarrow y = 2$

Put $y = 2$ in $x = 1 - y$; then $x = 1 - 2 = -1$; $(-1,2)$

EXAMPLE:

Solve: $-2x - 2y = -13; -4x + 2y = 10$

For the first equation above, you can add $-4x + 2y$ to the left side and 10 to the right side of the first equation: $-2x - 2y + (-4x + 2y) = -13 + 10$. Now, if you simplify, you get: $-2x - 2y - 4x + 2y = -3 \rightarrow -6x = -3 \rightarrow$ $x = 0.5$. Now, put 0.5 for the x in the first equation: $-2(0.5) - 2y = -13$. By solving this equation, $y = 6$

PRACTICES:

Solve each system of equation by substitution.

1) $-x + 5y = -4$ $x - 3y = 8$	2) $2x + 3y = -6$ $-2x - y = 8$
3) $x + 2y = -5$ $5x - 10y = 5$	4) $y = -x + 5$ $3x - y = -3$
5) $3x = 6$ $10y = 4x + 2$	6) $3x + 2y = 2$ $x + 4y = -6$
7) $4x + y = 3$ $2x + 4y = -2$	8) $4y = 2x + 3$ $x - 4y = -2$
9) $7y = 14x$ $2x - 5y = -24$	10) $5y = x + 2$ $3x - 12y = -5$

Score: ....

Answer Key

1) $(14, 2)$	2) $(-\frac{9}{2}, 1)$
3) $(-2, -\frac{3}{2})$	4) $(\frac{1}{2}, \frac{9}{2})$
5) $(2, 1)$	6) $(2, -2)$
7) $(1, -1)$	8) $(-1, \frac{1}{4})$
9) $(3, 6)$	10) $(-\frac{1}{3}, \frac{1}{3})$

Name: ....

Solving Systems of Equations by Elimination

- ✓ A system of equations has two equations and two variables. For example, look at the system of equations: $x - y = 1, x + y = 5$
- ✓ The simplest way to solve a system of equation is using the elimination method. The elimination method uses the addition property of equality. On each side of an equation, you can add the same value.

EXAMPLE:

What is the value of x *and* y in this system of equations? $\begin{cases} 3x - 4y = -20 \\ -x + 2y = 10 \end{cases}$

Solving Systems of Equations by Elimination: $\begin{array}{c} 3x - 4y = -20 \\ \underline{-x + 2y = 10} \end{array}$ ⇒ Multiply the second equation by 3, then add it to the first equation.

$\begin{array}{c} 3x - 4y = -20 \\ \underline{3(-x + 2y = 10)} \end{array} \Rightarrow \begin{array}{c} 3x - 4y = -20 \\ \underline{-3x + 6y = 30)} \end{array} \Rightarrow 2y = 10 \Rightarrow y = 5.$ Now, substitute 5 for y in the first equation and solve for x. $3x - 4(5) = -20 \rightarrow 3x - 20 = -20 \rightarrow x = 0$

PRACTICES:

Solve each system of equation by elimination.

1) $-5x + y = -5$ $\quad -y = -6x + 6$	2) $-6x - 2y = -2$ $\quad 2x - 3y = 8$
3) $5x - 4y = 8$ $\quad -6x + y = -21$	4) $10x - 4y = -24$ $\quad -x - 20y = -18$
5) $25x + 3y = -13$ $\quad 12x - 6y = -36$	6) $x - 8y = -7$ $\quad 6x + 4y = 10$
7) $-6x + 16y = 4$ $\quad 5x + y = 11$	8) $2x - 3y = -10$ $\quad 4x + 6y = -20$
9) $x - 5y = -8$ $\quad 3x + 7y = -2$	10) $x - 2y = -3$ $\quad 2x + 6y = -1$

Score: ....

Answer Key	
1) $(1,0)$	2) $(1,-2)$
3) $(4,3)$	4) $(-2,1)$
5) $(-1,4)$	6) $(1,1)$
7) $(2,1)$	8) $(-5,0)$
9) $(-3,1)$	10) $(-2,\frac{1}{2})$

Name: ....

Systems of Equations Word Problems

✓ Define your variables, write the system of equations then use elimination method for solving systems of equations.

EXAMPLE:

Tickets to a movie cost $5 for students and $8 for adults. Some friends purchased **20** tickets for **$115.00.** How many adults ticket did they buy?

Let x be the number of adult tickets and y be the number of student tickets. There are 20 tickets. Then: $x + y = 20$. The cost of students' ticket is $5 and for adults it is $8, and the total cost is $115. So, $8x + 5y = 20$. Now, we have a system of equations:

$$\begin{cases} x + y = 20 \\ 8x + 5y = 115 \end{cases}$$

Multiply the first equation by -5 and add to the second equation: $-5(x + y = 20) =$

$-5x - 5y = -100$

$8x + 5y + (-5x - 5y) = 115 - 100 \rightarrow 3x = 15 \rightarrow x = 5 \rightarrow 5 + y = 20 \rightarrow y = 15$.

There are 5 adults' tickets and 15 student tickets.

PRACTICES:

Solve.

1) A school of 220 students went on a field trip. They took 20 vehicles, some vans, and some minibuses. Find the number of vans and the number of minibuses they took if each van holds 5 students and each minibus hold 15 students.

2) The sum of two numbers is 28. Their difference is 12. Find the numbers.

3) A farmhouse shelters 20 animals, some are pigs, and some are gooses. Altogether there are 64 legs. How many of each animal are there?

4) The sum of the digits of a certain two–digit number is 15. Reversing it's increasing the number by 9. What is the number?

5) The difference of two numbers is 16. Their sum is 32. Find the numbers.

6) Tickets to a movie cost $5 for adults and $3 for students. A group of friends purchased 18 tickets for $82.00. How many adults ticket did they buy?

7) At a store, Eva bought two shirts and five hats for $154.00. Nicole bought three same shirts and four same hats for $168.00. What is the price of each shirt?

8) A farmhouse shelters 10 animals, some are pigs, and some are ducks. Altogether there are 36 legs. How many pigs are there?

9) A class of 195 students went on a field trip. They took 19 vehicles, some cars and some buses. If each car holds 5 students and each bus hold 25 students, how many buses did they take?

10) The sum of two numbers is 50. Their difference is 20. Find the numbers.

Score: ....

Answer Key

1) There are 8 van and 12 minibuses.	2) 8 and 20
3) There are 12 pigs and 8 gooses.	4) 78
5) 24 and 8.	6) 14
7) $32	8) 8
9) 5	10) 15 and 35

Name: ....

Linear Equations

- ✓ The equation of a line: $y = mx + b$
- ✓ Find the slope.
- ✓ Find the y-intercept. This can be done by substituting the slope and the coordinates of a point (x, y) on the line.

EXAMPLE:

What will be the equation of the line that passes through $(2, -2)$ and has a slope of 7?

$y = mx + b$ is the general slope-intercept form of the equation; where, m is the slope and b is the y-intercept.

By substitution of the given point and given slope, we have: $-2 = (2)(7) + b$

So, $b = -2 - 14 = -16$, and our required equation is $y = 7x - 16$.

PRACTICES:

Find the slope of the line through each pair of points.	Write the slope–intercept form of the equation of the line through the given points.
1) $(3, 1), (2, 4)$	2) Through: $(2,3), (4,2)$
3) $(-3, 4), (-1, 6)$	4) Through: $(8, -3), (6,7)$
5) $(4, 4), (6, -6)$	6) Through: $(0.5, 4), (2.5, 4.4)$
7) $(-1, 8), (5, -4)$	8) Through: $(4, -2), (2.5, 1)$
9) $(12, -3), (7, -3)$	10) Through: $(-1, 0.7), (-2.3, 2)$

Score: ....

Answer Key	
1) -3	2) $y = -\frac{1}{2}x + 4$
3) 1	4) $y = -5x + 37$
5) -5	6) $y = \frac{1}{5}x + \frac{39}{10}$
7) -2	8) $y = -2x + 6$
9) 0	10) $y = -x - 0.3$

Name: ....

Graphing Lines of Equations

✓ The equation of the line is: $y = mx + c$

If slope-intercept form of a line given the slope m and the y-intercept (the intersection of the line and y-axis)

EXAMPLE:

Sketch the graph of $y = 8x - 3$.

To graph this line, two points are required.

We have value of y is -3 when x is 0.

The value of x is $3/8$ when y is 0.

$x = 0 \rightarrow y = 8(0) - 3 = -3$,

$y = 0 \rightarrow 0 = 8x - 3 \rightarrow x = \dfrac{3}{8}$

Now, the two points are: $(0, -3)$ and $(\dfrac{3}{8}, 0)$.

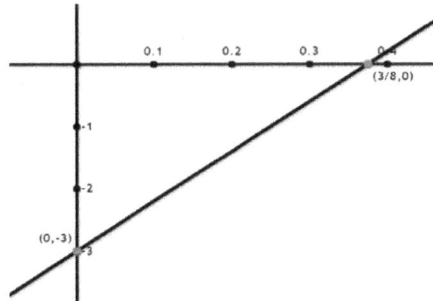

Find the points and graph the line. The slope of the line is 8.

PRACTICES:

Sketch the graph of each line.

1) $y = 3x - 2$

2) $y = 2x + 3$

3) $-2x = y + 5$

4) $4x + y = 2$

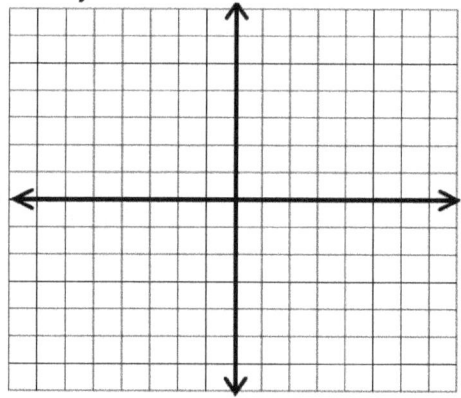

Score: ....

Answer Key

1)

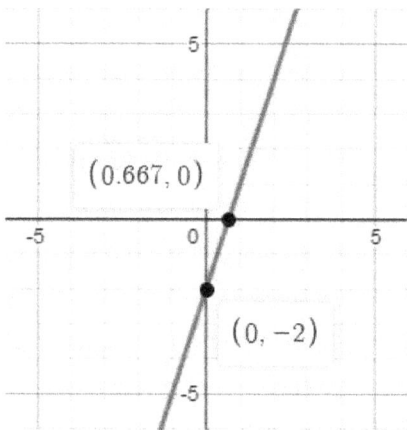

$(0.667, 0)$

$(0, -2)$

2)

3)

4)

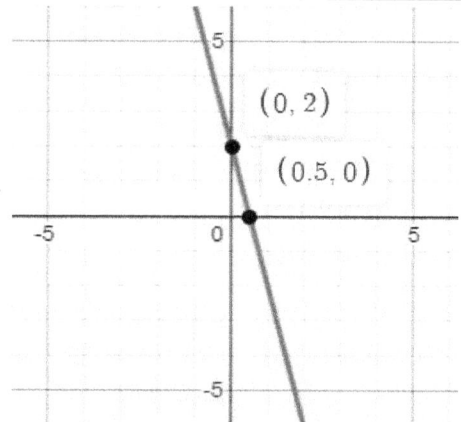

$(0, 2)$

$(0.5, 0)$

Name: ....

Graphing Linear Inequalities

- ✓ First step is to graph the "equals" line.
- ✓ Choose a testing point. (It can be any point on both sides of the line.)
- ✓ Put the value of (x, y) of that point in the inequality. If this satisfy the inequality, then this part of line is solution. If it does not satisfy then other part of line, is solution.

EXAMPLE:

Plot the graph of $y < 2x - 3$.

We know that first step is to graph the line: $y = 2x - 3$.

Now y-intercept is -3 and slope is 2. Then, select a

testing point. The simplest point to test is the origin:$(0, 0)$

$(0,0) \rightarrow y < 2x - 3 \rightarrow 0 < 2(0) - 3 \rightarrow 0 < -3$

0 is greater than -3. So, the other part of the line

(On the right side) is the solution.

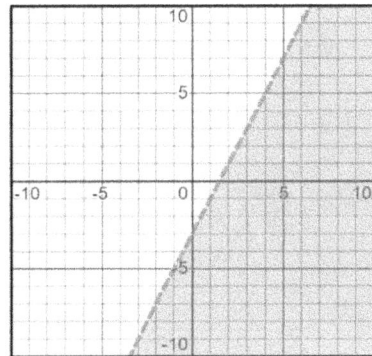

PRACTICES:

Sketch the graph of each linear inequality.

1) $2y + 8x \geq 4$

2) $-x + 2y \leq 4$

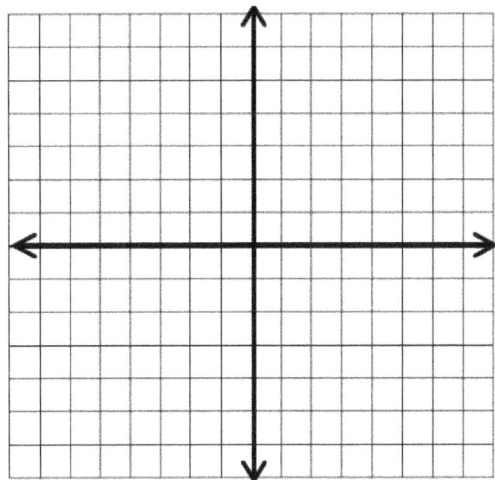

3) $2x + \frac{1}{2}y < 2$

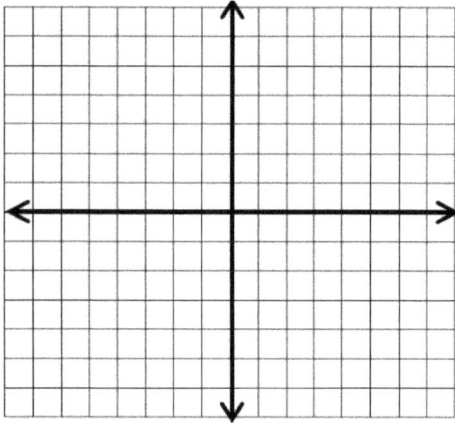

4) $-\frac{1}{3}x + y < 2$

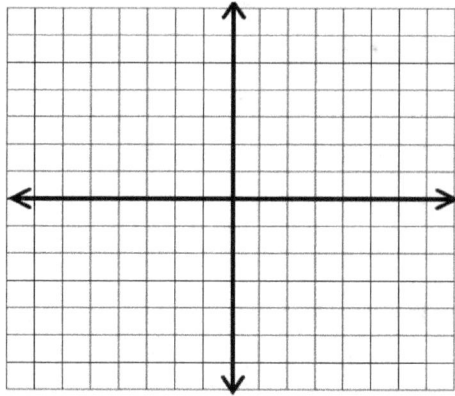

Score: ....

Answer Key

1)

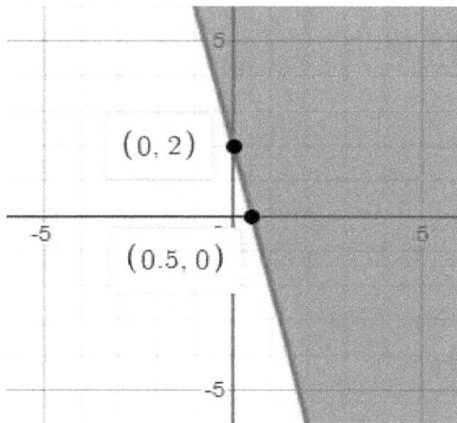

$(0, 2)$

$(0.5, 0)$

2)

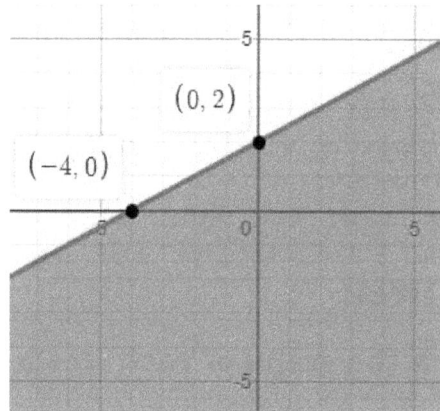

$(0, 2)$

$(-4, 0)$

3)

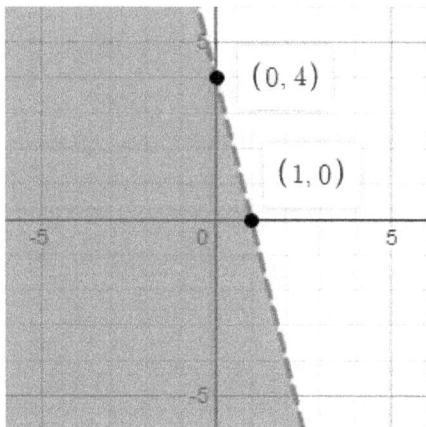

$(0, 4)$

$(1, 0)$

4)

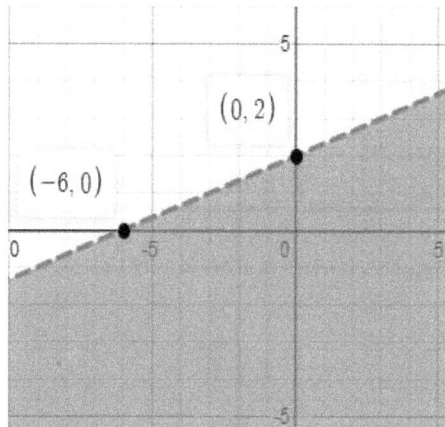

$(0, 2)$

$(-6, 0)$

Name: ....

Finding Distance of Two Points

- ✓ Distance of wo points A (x_1, y_1) and B (x_2, y_2): $d = \sqrt{(x_1 - x_2)^2 + (y_1 - y_2)^2}$
- ✓ The midpoint is middle of a line segment.
- ✓ We can find Midpoint of two endpoints A (x_1, y_1) and B (x_2, y_2) using this formula: $M\left(\frac{x_1+x_2}{2}, \frac{y_1+y_2}{2}\right)$

EXAMPLE:

Find the distance between of $(0, 8), (-4, 5)$.

Use distance of two points formula: $d = \sqrt{(x_1 - x_2)^2 + (y_1 - y_2)^2}$

$(x_1, y_1) = (0, 8)$ and $(x_2, y_2) = (-4, 5)$. Then: $d = \sqrt{(x_1 - x_2)^2 + (y_1 - y_2)^2} \rightarrow$

$d = \sqrt{(0 - (-4))^2 + (8 - 5)^2} = \sqrt{(4)^2 + (3)^2} = \sqrt{16 + 9} = \sqrt{25} = 5 \rightarrow d = 5$

PRACTICES:

Find the midpoint of the line segment with the given endpoints.	Find the distance between each pair of points.
1) $(1.5, -1), (0.5, -1)$	2) $(3, 4), (2, -1)$
3) $(1.5, -1), (0.5, -1)$	4) $(6, -1), (2, 3)$
5) $(0, 3), (4, -9)$	6) $(2, 5), (-2, 5)$
7) $(5, 2), (1, 5)$	8) $(0, -4), (-5, 1)$
9) $(-2, 0), (3, -4)$	10) $(3, -2), (-1, -5)$

Score: ....

Answer Key

1) $(1, -1)$	2) 5.09
3) $(1.5, -2)$	4) 5.656
5) $(2, -3)$	6) 4
7) $(3, 3.5)$	8) 7.07
9) $(0.5, -2)$	10) 5

Chapter 7 : Polynomials

Topics that you'll learn in this chapter:

- ➢ Classifying Polynomials

- ➢ Writing Polynomials in Standard Form

- ➢ Simplifying Polynomials

- ➢ Adding and Subtracting Polynomials

- ➢ Multiplying and Dividing Monomials

- ➢ Multiplying a Polynomial and a Monomial

- ➢ Multiplying Binomials

- ➢ Factoring Trinomials

- ➢ Operations with Polynomials

"Mathematics – the unshaken Foundation of Sciences, and the plentiful Fountain of Advantage to human affairs." — Isaac Barrow

Name: ....

Classifying Polynomials

Name	Degree	Example
constant	0	4
linear	1	$2x$
quadratic	2	$x^2 + 5x + 6$
cubic	3	$x^3 - x^2 + 4x + 8$
quartic	4	$x^4 + 3x^3 - x^2 + 2x + 6$
quantic	5	$x^5 - 2x^4 + x^3 - x^2 + x + 10$

EXAMPLE:

$17x^5 \Rightarrow$ Quantic binomial

PRACTICES:

Name each polynomial by degree and number of terms.	Write each polynomial in standard form
1) -5	2) $12x^4 + x - 4x^3$
3) $x + 1$	4) $12 - x^3 - 3x^5 + 9x^4$
5) $8x^6 - 7$	6) $x^2 + 13x^5 + x^3 - 4x$
7) $3x^2 - x$	8) $x^5 + 2x^3 (x^2 + 2)$
9) $-8x^4 + 3x^3 - 2x^2 - 3x$	10) $(x - 5)(x + 5)$

Score: ....

Answer Key	
1) Constant monomial	2) $12x^4 - 4x^3 + x$
3) Linear binomial	4) $-3x^5 + 9x^4 - x^3 + 12$
5) Sixth degree binomial	6) $13x^5 + x^3 + x^2 - 4x$
7) Quadratic binomial	8) $2x^6 + x^5 + 4x^3$
9) Quartic polynomial with four terms	10) $x^2 - 25$

Name: ....

Adding and Subtracting Polynomials

- ✓ To add polynomials, we combine like terms with some order of operations considerations thrown in.
- ✓ You have to be careful with a negative sign and don't confuse it with addition and subtraction.

EXAMPLE:

Add expressions. $(2x^3 - 6) + (9x^3 - 4x^2) = ?$

Remove parentheses: $(2x^3 - 6) + (9x^3 - 4x^2) = 2x^3 - 6 + 9x^3 - 4x^2$

Now combine like terms: $2x^3 - 6 + 9x^3 - 4x^2 = 11x^3 - 4x^2 - 6$

PRACTICES:

Simplify each expression.

1) $(x^3 + 6) - (6 + 3x^3)$	2) $(x^2 + 8) + (7x^2 - 8)$
3) $(2x^2 + x^3) - (5x^2 + 1)$	4) $(6x^2 - 4x) + (3x - 6x^2 + 1)$
5) $(x - 2x^3) - (4x^3 + 4)$	6) $(2x^3 + 2x^2) - (2x^2 - x^3 + 2)$
7) $(4x^2 - 3) + (x^2 - x^3)$	8) $(x^3 + 13x^4) - (13x^4 + 3x^3)$
9) $(x^4 + 2x^5 + 3x^3) + (4x^3 + 6x^4)$	10) $(3x^3 - 6x^6) + (3x^3 + 2x^6)$

Score: ....

Answer Key	
1) $-2x^3$	2) $8x^2$
3) $x^3 - 3x^2 + 1$	4) $-x + 1$
5) $-6x^3 + x - 4$	6) $3x^3 - 2$
7) $-x^3 + 5x^2 - 3$	8) $-2x^3$
9) $2x^5 + 7x^4 + 7x^3$	10) $-4x^6 + 6x^3$

Name: ....

Multiply and Divide Monomials

- ✓ When dividing monomials, we divide coefficients first and then divide their variables.
- ✓ If we have exponents with the same base, we'll subtract their powers.
- ✓ Exponent's rules:

$$x^a \times x^b = x^{a+b}, \qquad \frac{x^a}{x^b} = x^{a-b}$$

$$\frac{1}{x^b} = x^{-b}, \qquad (x^a)^b = x^{a \times b}$$

$$(xy)^a = x^a \times y^a$$

EXAMPLE:

Multiply expressions. $(-3x^7)(4x^3) = ?$

Use this formula: $x^a \times x^b = x^{a+b} \rightarrow x^7 \times x^3 = x^{10}$; Then: $(-3x^7)(4x^3) = -12x^{10}$

Dividing expressions. $\frac{18x^2 y^5}{2xy^4} = ?$

Use this formula: $\frac{x^a}{x^b} = x^{a-b}$, $\frac{x^2}{x} = x^{2-1} = x$ and $\frac{y^5}{y^4} = y^{5-4} = y$; Then: $\frac{18x^2 y^5}{2xy^4} = 9xy$

PRACTICES:

Simplify.

1) $(x^3 y^2)(42y^4)$	2) $\frac{100x^5 y^6}{25x^6 y^{11}}$
3) $(8x^4)(12x^5)$	4) $\frac{75x^{16} y^{10}}{5x^6 y^7}$
5) $(-2x^{-3} y^2)^2$	6) $\frac{15x^{12} y^5}{5x^9 y^2}$
7) $(11x^2 y^4)(4x^9 y^{10})$	8) $\frac{50x^4 y^7}{25x^3 y^7}$
9) $(2x^{-3} y^4)^2$	10) $\frac{-21x^8 y^{13}}{3x^6 y^6}$

Score: ....

Answer Key

1) $42x^3y^6$	2) $4x^{-1}y^{-5}$
3) $96x^9$	4) $15x^{10}y^3$
5) $4x^{-6}y^4$	6) $3x^3y^3$
7) $44x^{11}y^{14}$	8) $2x$
9) $4x^{-6}y^8$	10) $-7x^2y^7$

Name: ....

Multiplying Monomials

✓ A polynomial having only one term is called monomial, like $2x$ or $7y$

EXAMPLE:

Multiply expressions. $5a^4b^3 \times 2a^3b^2 = ?$

Use this formula: $x^a \times x^b = x^{a+b}$

$a^4 \times a^3 = a^{4+3} = a^7$ and $b^3 \times b^2 = b^{3+2} = b^5$

Then: $5a^4b^3 \times 2a^3b^2 = 10a^7b^5$

PRACTICES:

Simplify each expression.

1) $2xy^2 \times 3z^2$

2) $3xyz \times 5x^2y$

3) $4pq^3 \times (-3p^3q)$

4) $s^3t^2 \times 2st^5$

5) $5p^3 \times (-2p^2)$

6) $-2p^2r \times 6pr^3$

7) $(-a)(-4a^6b)$

8) $2u^2v^3 \times (-8u^3v^3)$

9) $6u^3 \times (2u)$

10) $-5y^2 \times 4x^2y$

Score: ....

Answer Key	
1) $6xy^2z^2$	2) $15x^3y^2z$
3) $-12p^4q^4$	4) $2s^4t^{10}$
5) $-10p^5$	6) $-12p^3r^4$
7) $4a^7b$	8) $-16u^5v^6$
9) $12u^4$	10) $-20x^2y^3$

Name: ....

Multiply a Polynomial and a Monomial

- ✓ Use the product rule for exponents to multiply monomials.
- ✓ Use distributive property to multiply a monomial by a polynomial.

$$a \times (b + c) = a \times b + a \times c$$

EXAMPLE:

Multiply expressions. $-4x(5x + 9) = ?$

Use Distributive Property: $-4x(5x + 9) = -20x^2 - 36x$

PRACTICES:

Find each product.

1) $3(2x - 2y)$	2) $5x(4x - y)$
3) $-2x(x + 5)$	4) $11(3x + 7)$
5) $10x(5x - 2y)$	6) $4(3x - 5y)$
7) $2x(3x^3 - 5x + 4)$	8) $-4x (2 + 4xy)$
9) $3(2x^2 - 8x + 3)$	10) $-3x^2(3x^2 + 5)$

Score: ....

Answer Key

1) $6x - 6y$	2) $20x^2 - 5xy$
3) $-2x^2 - 10$	4) $33x + 77$
5) $50x^2 - 20xy$	6) $50x^2 - 20xy$
7) $6x^4 - 10x^2 + 8x$	8) $-16x^2y - 8x$
9) $6x^2 - 24x + 9$	10) $-9x^4 - 15x^2$

Name: ....

Multiply Binomials

✓ Use "FOIL". (First–Out–In–Last)

$$(x + a)(x + b) = x^2 + (b + a)x + ab$$

EXAMPLE:

Multiply Binomials. $(x + 5)(x - 2) =?$

Use "FOIL". (First–Out–In–Last):

$(x + 5)(x - 2) = x^2 - 2x + 5x - 10$
Then simplify: $x^2 - 2x + 5x - 10 = x^2 + 3x - 10$

PRACTICES:

Multiply.

1) $(2x - 2)(x + 3)$	2) $(4x + 2)(2x + 1)$
3) $(x + 3)(x + 4)$	4) $(x^2 + 5)(x^2 - 5)$
5) $(2x - 3)(x + 4)$	6) $(2x - 6)(x + 7)$
7) $(x - 2)(3x - 4)$	8) $(2x - 5)(x + 4)$
9) $(x + 10)(x - 10)$	10) $(x - 3)(3x + 4)$

Score: ....

Answer Key

1) $2x^2 + 4x - 6$	2) $8x^2 + 8x + 2$
3) $x^2 + 7x + 12$	4) $x^4 - 25$
5) $2x^2 + 5x - 12$	6) $2x^2 + 8x - 42$
7) $3x^2 - 10x + 8$	8) $2x^2 + 3x - 20$
9) $x^2 - 100$	10) $3x^2 - 5x - 12$

Factor Trinomials

- ✓ FOIL":

$$(x + a)(x + b) = x^2 + (b + a)x + ab$$

- ✓ "Difference of Squares":

$$a^2 - b^2 = (a + b)(a - b)$$
$$a^2 + 2ab + b^2 = (a + b)(a + b)$$
$$a^2 - 2ab + b^2 = (a - b)(a - b)$$

- ✓ "Reverse FOIL":

$$x^2 + (b + a)x + ab = (x + a)(x + b)$$

EXAMPLE:

Factor this trinomial. $x^2 - 2x - 8 = ?$

Expression is broken into groups: $(x^2 + 2x) + (-4x - 8)$

Now x is a factor from $x^2 + 2x : x(x + 2)$ and factor out -4 from $-4x - 8: -4(x + 2)$

Then: $= x(x + 2) - 4(x + 2)$, now factor out like term: $x + 2$

Then: $(x + 2)(x - 4)$

PRACTICES:

Factor each trinomial.

1) $x^2 - 12x + 27$	2) $x^2 + 5x - 24$
3) $x^2 + 13x + 30$	4) $x^2 - 81$
5) $2x^2 + 12x - 14$	6) $x^2 + 2x - 8$
7) $2x^2 + 3x + 1$	8) $2x^2 + 2x - 4$
9) $9x^2 + 3x - 2$	10) $x^2 + 15x + 56$

Score: ....

Answer Key

1) $(x - 3)(x - 9)$	2) $(x + 8)(x - 3)$
3) $(x + 10)(x + 3)$	4) $(x + 9)(x - 9)$
5) $(x + 7)(2x - 2)$	6) $(x - 2)(x + 4)$
7) $(2x + 1)(x + 1)$	8) $(2x - 2)(x + 2)$
9) $(3x - 1)(3x + 2)$	10) $(x + 7)(x + 8)$

Name: ....

Operations with Polynomials

✓ Use distributive property to multiply a monomial by a polynomial.

$$a \times (b + c) = a \times b + a \times c$$

EXAMPLE:

Multiply. $5(2x - 6) =$

Use the distributive property: $5(2x - 6) = 5 \times 2x - 5 \times (-6) = 10x - 30$

PRACTICES:

Find each product.

1) $x^2(3x - 2)$	2) $2x^2(5x - 3)$
3) $-x(5x - 3)$	4) $x^2(-3x + 9)$
5) $5(7x + 3)$	6) $8(3x + 8)$
7) $5(10x + 4)$	8) $-3x^5(x - 3)$
9) $5(3x^2 - x + 2)$	10) $4(x^2 - 2x + 3)$

Score: ....

Answer Key	
1) $3x^3 - 2x^2$	2) $10x^3 - 6x^2$
3) $-5x^2 + 3x$	4) $-3x^3 + 9x^2$
5) $35x + 15$	6) $24x + 64$
7) $50x + 20$	8) $-3x^6 + 9x^5$
9) $15x^2 - 5x + 10$	10) $4x^2 - 8x + 12$

Name: ....

Simplifying Polynomials

- ✓ Find "identical" terms. (They have same variables with same power).
- ✓ Use "FOIL". (First-Out-In-Last) for binomials:

$$(x + a)(x + b) = x^2 + (b + a)x + ab$$

- ✓ Using order of operation, add or subtract "identical" terms

EXAMPLE:

Simplify this expression. $(4 + x)(x - 3) =$?

Use FOIL: $(x + 4)(x - 3) = x^2 + x - 12$

PRACTICES:

Simplify each expression.

1) $-3x^2 + x^5 + 7x^5 - 2x^2 + 6$	2) $18x^5 - 3x^5 + 7x^2 - 15x^5 + 4$
3) $x(x^3 + 9) - 6(8 + x^2)$	4) $x(x^2 + 2x^3) - x^3 + x$
5) $4 - 17x^2 + 30x^2 - 17x^2 + 26$	6) $4x^2 - 8x + 3x^3 + 15x - 20x$
7) $(x - 6)(x - x^2 + 5)$	8) $(x - 5)(x + 5)$
9) $(x^4 - x) + (4x^2 - 3x^4)$	10) $x(x^2 + x + 3)$

Score: ....

Answer Key	
1) $8x^5 - 5x^2 + 6$	2) $7x^2 + 4$
3) $x^4 - 6x^2 + 9x - 48$	4) $2x^4 + x$
5) $-4x^2 + 30$	6) $3x^3 + 4x^2 - 13x$
7) $-x^3 + 7x^2 - x - 30$	8) $x^2 - 25$
9) $-2x^4 + 4x^2 - x$	10) $x^3 + x^2 + 3x$

Chapter 8 : Functions

Topics that you'll learn in this chapter:

- ➤ Relations and Functions

- ➤ Rate of change and Slope

- ➤ x and y intercept

- ➤ Slope-intercept form

- ➤ Slope-point form

- ➤ Equation of Parallel or Perpendicular lines

- ➤ Equation of Horizontal and Vertical Lines

- ➤ Function Notation

- ➤ Adding and Subtracting Functions

- ➤ Multiplying and Dividing Functions

- ➤ Composition of Functions

- ➤ Solve a Quadratic Equation

It's fine to work on any problem, so long as it generates interesting mathematics along the way – even if you don't solve it at the end of the day." – Andrew Wiles

Name:

....

Relations and Functions

✓ **RELATION:** A relationship between two sets of elements like input and output, input can have as many as outputs.

✓ **FUNCTION:** A relationship between two sets of elements like input and output and only and exactly one output related to one input.

✓ A function is a type of relation, but a relation may not be a function.

✓ **The Vertical Line Test**

A graphical method that we can observe the cross between the graph and vertical line.

The graph is a function if and only if the intersection point is one.

EXAMPLE:

Use the vertical line test to determine if the graph is function or not?

The graph is not a function

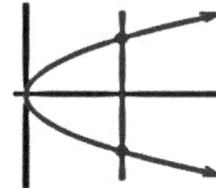

PRACTICES:

State the domain and range of each relation. Then determine whether each relation is a function.

1)

Function:

....

Domain:

....

Range:

....

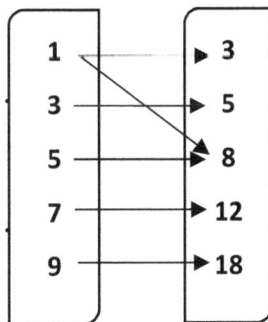

2)

Function:

....

Domain:

....

Range:

....

3) $\{(1,-2),(4,-1),(0,5),(4,0),(3,8)\}$	4)		x	y
Function:	Function:		3	4
....	....		0	1
Domain:	Domain:		-2	-3
....	....		6	-3
Range:	Range:		8	2
....	....			

Score: ....

Answer Key

1) No, $D_f = \{1,3,5,7,9\}$, $R_f = \{3,5,8,12,18\}$	2) Yes, $D_f = (-\infty,\infty)$, $R_f = \{2,-\infty)$
3) No, $D_f = \{1,4,0,3\}$, $R_f = \{-2,-1,5,0,8\}$	4) Yes, $D_f = \{3,0,-2,6,8\}$, $R_f = \{4,1,-3,2\}$

Name: ....

Rate of change

- ✓ Slope can be described as "rate of change".

- ✓ Rate of change is a ratio between a change in one variable comparing to a corresponding change in another variable. Rate of change $= \frac{change\ in\ output\ (y)}{change\ in\ input\ (x)}$

- ✓ Rates of change can be positive, negative, or zero.

EXAMPLE:

The table shows the amount of money SB carwash made washing car. Find the rate of change in dollar per car?

SB Carwash	Number	4	8	12	16
	Money ($)	32	56	80	104

Rate of change $= \frac{change\ in\ output\ (y)}{change\ in\ input\ (x)} = \frac{Change\ in\ money}{Change\ in\ car} = \frac{56-32}{8-4} = \frac{24}{4} = \frac{6}{1}$, or $6 per car

PRACTICES:

What is the average rate of change of the function?

1)

Gallons	3	5	7	9
Miles	81	135	189	243

2)

Products	145	159	173	187
Costs	761	719	677	635

3)

x	4.5	6	7.5	9
y	6	15	24	33

4)

x	41	47	53	59
y	67	52	37	22

5) $f(x) = -2x + 4$, from $x = -1$ to $x = 4$?

6) $f(x) = x - 6$, from $x = -5$ to $x = 1$?

7) $f(x) = -4$, from $x = 3$ to $x = -2$?

8) $f(x) = 3x^2 + 5$, from $x = 3$ to $x = 6$?

9) $f(x) = -2x^2 - 4$, from $x = 2$ to $x = 4$?

10) $f(x) = x^3 + 3$, from $x = 1$ to $x = 2$?

Score: ....

Answer Key	
1) 27 miles per gallon	2) -3 cost per product
3) 6	4) -2.5
5) -2	6) 1
7) 0	8) 27
9) -12	10) 7

Name: ....

Slope

- ✓ The slope is used to describe the steepness and direction of lines on the coordinate plane.
- ✓ A coordinate plane is a two-dimensional plane formed by the intersection of a vertical line called y-axis and a horizontal line called x-axis. These are perpendicular lines that intersect each other at zero, and this point is called the origin.
- ✓ An ordered pair (x, y) shows the location of a point.
- ✓ A line on coordinate plane can be drawn by connecting two points.
- ✓ The slope of a line with two points A (x_1, y_1) and B (x_2, y_2) can be found by using this formula: $\frac{y_2 - y_1}{x_2 - x_1} = \frac{rise}{run}$

EXAMPLE:

Use the given points to determine the slope. Points, $(3, -7)$, $(-2, -9)$

$$\text{Slope} = \frac{y_2 - y_1}{x_2 - x_1} = \frac{-9 - (-7)}{-2 + 3} = \frac{-9 + 7}{1} = \frac{-2}{1} = -2$$

PRACTICES:

Find the slope of the line through each pair of points.

1) $(2, -9), (5, -6)$	2) $(14, 7), (20, 12)$
3) $(1, -5), (8, -4)$	4) $(13, -9), (15, -7)$
5) $(-5, -8), (-8, -2)$	6) $(0, 0), (12, -2)$
7) $(14, -8), (-6, 5)$	8) $(-2, 5), (-2, 8)$
9) $(-14, -9), (-6, -15)$	10) $(-19, 2), (2, -19)$

Score: ....

Answer Key

1) 1	2) $\frac{5}{6}$
3) $\frac{1}{7}$	4) 1
5) -2	6) $-\frac{1}{6}$
7) $-\frac{13}{20}$	8) Undefined
9) $-\frac{3}{4}$	10) -1

Name: ....

x and y intercept

✓ x-intercept is the point at which the graph crosses the x-axis, and the value of y is zero.

✓ y-intercept is the point at which the graph crosses the y-axis, and the value of x is zero.

EXAMPLE:

Find the intercepts of the equation y = 3x − 21.

To find the x-intercept, set $y = 0$, then $0 = 3x − 21 \rightarrow 3x = 21 \rightarrow x = 7$. x-intercept(7,0)

To find the y-intercept, set $y = 0$, then y = 3(0) − 21 → y = −21. y-intercept (0, −12)

PRACTICES:

Find the x and y intercepts for the following equations.

1) $5x + 3y = 15$	2) $y = x + 8$
3) $4x = y + 16$	4) $x + y = -2$
5) $4x - 3y = 7$	6) $7y - 5x + 10 = 0$
7) $\frac{3}{7}x + \frac{1}{4}y + \frac{2}{3} = 0$	8) $3x - 21 = 0$
9) $24 - 4y = 0$	10) $-2x - 6y + 42 = 12$

Score: ....

Answer Key

1) $y - intercept = 5$
   $x - intercept = 3$

2) $y - intercept = 8$
   $x - intercept = -8$

3) $y - intercept = -16$
   $x - intercept = 4$

4) $y - intercept = -2$
   $x - intercept = -2$

5) $y - intercept = -\frac{7}{3}$
   $x - intercept = \frac{7}{4}$

6) $y - intercept = -\frac{10}{7}$
   $x - intercept = 2$

7) $y - intercept = -\frac{8}{3}$
   $x - intercept = -\frac{2}{7}$

8) $y - intercept = undefind$
   $x - intercept = 7$

9) $y - intercept = 6$
   $x - intercept = undefind$

10) $y - intercept = 5$
    $x - intercept = 15$

Name: ....

Writing Linear Equations

✓ The equation of a line:
$$y = mx + b$$
✓ Identify the slope.
✓ Find the y-intercept. This can be done by substituting the slope and the coordinates of a point (x, y) on the line.

EXAMPLE:

If f is a linear function, with $f(4) = -2$, and $f(7) = 1$, find an equation for the function.

Slope: $\frac{y_2 - y_1}{x_2 - x_1} = \frac{1 - (-2)}{7 - 4} = \frac{1 + 2}{3} = 1$

$y = mx + b \rightarrow y = 1 \times x + b \rightarrow y = x + b$, and $f(7) = 1$, then $1 = 7 + b \rightarrow b = -6$

The linear equation is: $y = x - 6$

PRACTICES:

Write the equation of each line in form of $y = mx + b$.

1) $m = 2$; y-intercept= -7	2) $m = -\frac{2}{5}$; y-intercept= $\frac{2}{3}$
3) $f(-2) = 1$; $f(-3) = 4$	4) $f(6) = -1$; $f(2) = 7$
5) Through: $(5, 7), (3, 6)$	6) Through: $(-1.5, 2), (6.5, -2)$
7) Through: $(2, -1), (6, 11)$	8) Through: $(4, 1), (-2, 7)$
9) Through: $(2, 4), (-2, -4)$	10) Through: $(4, 5), (0, -1)$

Score: ....

	Answer Key	
1) $y = 2x - 7$		2) $y = -\frac{2}{5}x + \frac{2}{3}$
3) $y = -3x - 5$		4) $y = -2x + 11$
5) $y = \frac{1}{2}x + \frac{9}{2}$		6) $y = -0.5x + 1.25$
7) $y = 3x - 7$		8) $y = -x + 5$
9) $y = 2x$		10) $y = 1.5x - 1$

Name: ....

Slope-intercept form

- ✓ The slope-intercept form is one of several ways you can write the equation of a line.
- ✓ Using the slope m and the y-intercept b, then the equation of the line is:

$$y = mx + b$$

EXAMPLE:

Solve for y when 4x - 2y = 12.

Subtract $4x$ from both sides: $-4x + 4x - 2y = 12 - 4x \rightarrow -2y = 12 - 4x$

Divide everything by -2: $y = -6 + 2x \rightarrow y = 2x - 6$

PRACTICES:

Write the slope–intercept form of the equation of each line.

1) $y - 4 = x + 3$	2) $5x + 14 = -3y$
3) $18x - 12y = -6$	4) $7x - 4y + 25 = 0$
5) $-\frac{1}{3}y = -2x + 3$	6) $5 - y - 4x = 0$
7) $-y = -6x - 9$	8) $-2(7x + y) = 24$
9) $3(y + 3) = 2(x - 3)$	10) $\frac{3}{4}y + \frac{1}{4}x + \frac{5}{4} = 0$

Score: ....

Answer Key	
1) $y = x + 7$	2) $y = -\frac{5}{3}x - \frac{14}{3}$
3) $y = \frac{3}{2}x + \frac{1}{2}$	4) $y = \frac{7}{4}x + \frac{25}{4}$
5) $y = 6x - 9$	6) $y = -4x + 5$
7) $y = 6x + 9$	8) $y = -7x - 12$
9) $y = \frac{2}{3}x - 5$	10) $y = -\frac{1}{3}x - \frac{5}{3}$

Name: ....

Point-slope form

✓ Using the slope m and a point (x_1, y_1) on the line, the equation of the line is:

$$(y - y_1) = m(x - x_1)$$

EXAMPLE:

Write the point-slope form of an equation of a line with a slope of 3 that passes through the point $(3, -2)$.

The slope is 3, so m = 3. We also know one point, so we know $x_1 = 3$ and $y_1 = -2$. Now we can substitute these values into the general point-slope equation.

$$y - y_1 = m(x - x_1) \rightarrow y - (-2) = \frac{3}{8}(x - 3) \rightarrow y + 2 = \frac{3}{8}(x - 3)$$

PRACTICES:

Write an equation in point–slope form for the line that passes through the given point with the slope provided.

1) $(2, -3), m = 4$	2) $(-7, 4), m = \frac{1}{5}$
3) $(0, -6), m = -2$	4) $(-a, b), m = m$
5) $(-9, 1), m = 3$	6) $(3, 0), m = -5$
7) $(-4, 11), m = \frac{1}{3}$	8) $(0, 11), (-2, 11)$
9) $\left(-\frac{1}{3}, 3\right), m = \frac{1}{5}$	10) $(0, 0), (1, -3)$

Score: ....

Answer Key	
1) $y + 3 = 4(x - 2)$	2) $y - 4 = \frac{1}{5}(x + 7)$
3) $y + 6 = -2x$	4) $y - b = m(x + a)$
5) $y - 1 = 3(x + 9)$	6) $y = -5(x - 3)$
7) $y - 11 = \frac{1}{3}(x + 4)$	8) $y - 11 = 0$
9) $y - 3 = \frac{1}{5}\left(x + \frac{1}{3}\right)$	10) $y = -3x$

Name: ....

Equation of Parallel or Perpendicular lines

✓ Parallel lines: The two lines will never intersect, and their slopes are identical. The only difference between the two lines is the y-intercept.

$\begin{cases} y = m_1x + b_1 \\ y = m_2x + b_2 \end{cases}$; Then, $m_1 = m_2$ and $b_1 \neq b_2$.

✓ Perpendicular Lines do intersect. Their intersection forms a right, or 90° angle. The slope of one line is the negative reciprocal of the slope of the other line.

$\begin{cases} y = m_1x + b_1 \\ y = m_2x + b_2 \end{cases}$; then, $m_1 = -\frac{1}{m_2}$, or $m_1 \times m_2 = -1$

EXAMPLE:

Given the functions below, identify the functions whose graphs are a pair of parallel lines

and a pair of perpendicular lines. $\begin{cases} f(x) = 4x - 5 & g(x) = -4x + 2 \\ h(x) = \frac{1}{4}x - 5 & p(x) = 4x + 7 \end{cases}$

Parallel lines have the same slope ($f(x)$, and $p(x)$)

Perpendicular lines have negative reciprocal slopes ($g(x)$, and $h(x)$).

PRACTICES:

Write an equation of the line that passes through the given point and is parallel to the given line.	Write an equation of the line that passes through the given point and is perpendicular to the given line.
1) $(0,7), -5x - y = -4$	6) $(\frac{3}{5}, \frac{2}{5}), y = -6x - 24$
2) $(-2,-1), y = \frac{4}{5}x + 3$	7) $(-10, 0), y = \frac{5}{3}x - 15$
3) $(-2,5), -8x + 5y = -18$	8) $(3,-5), y = x + 12$
4) $(3,-2), y = -\frac{2}{5}x - 3$	9) $(-3,-1), y = \frac{7}{3}x - 4$
5) $(-5,-5), 6x + 15y = -30$	10) $(0,0), y - 8x + 6 = 0$

Score: ....

Answer Key	
1) $y = -5x + 7$	2) $y = \frac{4}{5}x + \frac{3}{5}$
3) $y = \frac{8}{5}x + \frac{41}{5}$	4) $y = -\frac{2}{5}x - \frac{4}{5}$
5) $y = -\frac{2}{5}x - 7$	6) $y = \frac{1}{6}x + \frac{3}{10}$
7) $y = -\frac{3}{5}x - 6$	8) $y = -x - 2$
9) $y = -\frac{3}{7}x - \frac{16}{7}$	10) $y = -\frac{1}{8}x$

Equation of Horizontal and Vertical Lines

- ✓ Equations of horizontal and vertical lines only have one variable.
- ✓ The slope of horizontal lines is 0 and y-values for each point are the same. Then, the equation of horizontal lines is: $y = b$.
- ✓ The slope of vertical lines is undefined and the equation for a vertical line is: $x = a$

EXAMPLE:

Write an equation for the vertical line that passes through $(4, -1)$.

As the line is vertical, x is constant, and x always takes the same value. Then x always takes the value 4. Thus, the equation is $x = 4$.

PRACTICES:

Sketch the graph of each line.

1) $y = 3$

2) $y = -1$

3) $x = 0$

4) $x = 3$

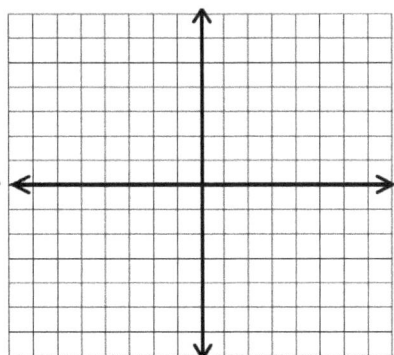

Score: ....

Answer Key

1) $y = 3$

2) $y = -1$

3) $x = 0$

4) $x = 3$

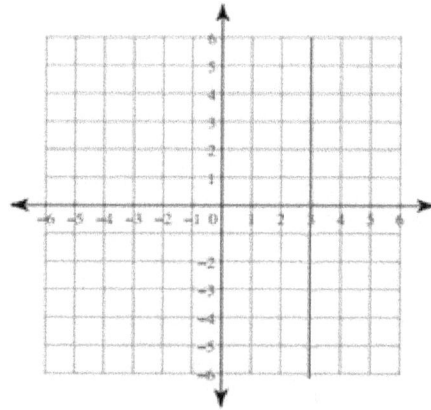

Name: ....

Function Notation

- ✓ **Function notation** is a way to write functions that is easy to read and understand. Also, it determines that a relationship is a function.
- ✓ The notation y = f (x) defines a function named f. This is read y is a function of x. The letter x is the input value or independent variable. The letter y, or f (x), is the output value or dependent variable.

EXAMPLE:

Write $y = x^2 + 3x - 5$ using function notation and evaluate the function at $x = 2$.

By applying function notation, we get: $f(x) = x^2 + 3x - 5$

Evaluation: Substitute x with 2:

$f(2) = 2^2 + 3 \times 2 - 5 = 4 + 6 - 5 = 5$

PRACTICES:

Write in function notation.

1) $v = 8t$	2) $r = 4p^2 + 2p - 2$
3) $h = 15g + 8$	4) $y = 5x - \dfrac{3}{4}$

Evaluate each function.

5) $h(x) = x^3 - 8$, find $h(-2)$	6) $f(u) = 9u - 2$, find $f(^1/_3)$
7) $h(x) = 3x - 6$, find $h(a)$	8) $h(a) = -2a + 4$, find $h(3b)$
9) $h(x) = x^2 + 4x - 7$, find $h(x^2)$	10) $h(x) = x^2 + 5$, find $h(-\dfrac{a}{3})$

Score: ....

Answer Key	
1) $v(t) = 8t$	2) $r(p) = 4p^2 + 2p - 2$
3) $h(g) = 15g + 8$	4) $f(x) = 5x - \frac{3}{4}$
5) -16	6) 1
7) $3a - 6$	8) $-6b + 4$
9) $x^4 + 4x^2 - 7$	10) $\frac{1}{9}a^2 + 5$

Name: ....

Adding and Subtracting Functions

✓ Like numbers and polynomials, we can add and subtract functions which results into a new function.

✓ Let f(x) and g(x) be two functions:

We can add two functions as: $(f + g)(x) = f(x) + g(x)$

We can subtract two functions as: $(f - g)(x) = f(x) - g(x)$

EXAMPLE:

Find the sum, and difference of $f = 3x + 1$ and $g = x - 2$ at the point 5.

$$f + g = (3x + 1) + (x - 2) = 4x - 1 \rightarrow (f + g)(5) = 19$$
$$f - g = (3x + 1) - (x - 2) = 2x + 3 \rightarrow (f - g)(5) = 13$$

PRACTICES:

Perform the indicated operation.

1) $h(t) = 5t - 3$ $g(t) = 5t + 3$ Find $(h - g)(t)$.	2) $h(n) = 4n - 4$ $g(n) = n^2 - 6n + 9$ Find $(h + g)(a)$.
3) $g(a) = -3a^2 + 4$ $f(a) = 2a^2 - a + 4$ Find $(g - f)(a)$.	4) $g(x) = -x^2 + 8 - 3x$ $f(x) = 8 + 2x$ Find $(g - f)(x)$.
5) $h(x) = 3x^2 - 5$ $g(x) = -4x^2 + 2x$ Find $(h + g)(t)$.	6) $g(t) = t + 8$ $f(t) = -3t^2 + t$ Find $(g - f)(u - 1)$.
7) $h(x) = -3x + 4$ $g(x) = 2x - 6$ Find $(h + g)(2)$.	8) $k(x) = -3x + 6$ $h(x) = x^2 + 2x + 4$ Find $(k + h)(t - 3)$.

Score: ....

Answer Key	
1) -6	2) $a^2 - 2a + 5$
3) $-5a^2 + a$	4) $-x^2 - 5x$
5) $-t^2 + 2t - 5$	6) $3u^2 - 6u + 11$
7) -4	8) $t^2 - 7t + 22$

Name: ....

Multiplying and Dividing Functions

- ✓ Like numbers and polynomials, we can multiply and divide functions which results into a new function.
- ✓ Let f(x) and g(x) be two functions:
  We can multiply two functions as: $(f \cdot g)(x) = f(x) \cdot g(x)$
  We can divide two functions as: $\left(\frac{f}{g}\right)(x) = \frac{f(x)}{g(x)}$

EXAMPLE:

Given that $f(x) = x + 3$ and $g(x) = x^2 - 9$, find $(fg)(x)$ and $\left(\frac{f}{g}\right)(x)$ at the point 1.

$(fg)(x) = f(x) \times g(x) = (x+3)(x^2-9) = x^3 + 3x^2 - 9x - 27 \rightarrow (fg)(1) = -32$

$\left(\frac{f}{g}\right)(x) = \frac{f(x)}{g(x)} = \frac{x+3}{x^2-9} = \frac{x+3}{(x-3)(x+3)} = \frac{1}{x-3} \rightarrow \left(\frac{f}{g}\right)(1) = -\frac{1}{2}$

PRACTICES:

Perform the indicated operation.

1) f(x) = $x^2 - 3x$ g(x) = $4x^2 - 2$ Find (f.g)(x)	2) g(t) = $\frac{1}{3}t^2 + \frac{1}{3}$ h(t) = 3t – 3 Find (h.g)($\frac{1}{3}$)
3) $f(a) = 12a - 8$ $g(a) = 5a + 10$ Find $\left(\frac{f}{g}\right)(-1)$	4) $h(a) = -2a$ $g(a) = -6a^2 - 2a$ Find $\left(\frac{h}{g}\right)(a)$
5) $g(a) = 4a - 3$ $h(a) = 4a - 2$ Find $(g.h)(2)$	6) $k(n) = 2n^2 - n$ $h(n) = 3n^2 - 3$ Find $(k.h)(1)$
7) $f(x) = 2x + 1$ $g(x) = 4x^2 - 1$ Find $\left(\frac{g}{f}\right)(x)$	8) $f(t) = -a + 3$ $g(t) = a^3 + 2$ Find $\left(\frac{3f}{g}\right)(a)$

Score: ....

Answer Key	
1) $4x^4 - 12x^3 - 2x^2 + 6x$	2) $-\dfrac{20}{27}$
3) -4	4) $\dfrac{1}{3a+1}$
5) 30	6) 0
7) $2x - 1$	8) $\dfrac{-3a+9}{a^3+2}$

GED Math Prep

Name: ....

Composition of Functions

- ✓ A composite function is generally a function that is written inside another function. Composition of a function is done by substituting one function into another function.
- ✓ The notation used for composition is:
$$(fog)(x) = f(g(x))$$

EXAMPLE:

Using $f(x) = x + 1$ and $g(x) = 2x$, find: $(f\ o\ g)(1)$

$(f\ o\ g)(x) = f(g(x)) = 2(x + 1) = 2x + 2$
$(f\ o\ g)(1) = 4$

PRACTICES:

Using $f(x) = 2x - 8$, and $g(x) = -2x + 1$, find:

1) $f(g(0))$

2) $f(f(1))$

3) $g(f(3))$

Using $f(x) = 3x - 4a$, and $g(x) = x^2 - 2$, find:

4) $(fog)(1) = f(g(1))$

5) $(fof)(3)$

6) $(gof)(2)$

Using $f(x) = -2x + 3$, and $g(x) = x - b$, find:

7) $(fog)(-2x)$

8) $(fog)(x + 1)$

9) $(gof)(x^2)$

Score: ....

Answer Key	
1) -6	2) -20
3) 5	4) $-3 - 4a$
5) $27 - 16a$	6) $16a^2 - 48a + 34$
7) $4x + 3 + 2b$	8) $-2x + 1 + 2b$
9) $-2x^2 + 3 - b$	

Solve a Quadratic Equation

✓ Write the equation in the Standard form:
$ax^2 + bx + c = 0$ (One side must only contain zero)
✓ Factorize the quadratic.
✓ Use quadratic formula if you couldn't factorize the quadratic.
✓ Quadratic formula: $x = \frac{-b \pm \sqrt{b^2 - 4ac}}{2a}$

EXAMPLE:

Solve: $x^2 - 4x - 21 = 0$.

$\begin{cases} a = 1 \\ b = -4 \\ c = -21 \end{cases} \Rightarrow x = \frac{-b \pm \sqrt{b^2 - 4ac}}{2a} = \frac{-(-4) \pm \sqrt{(-4)^2 - 4(1)(-21)}}{2(1)} = \begin{cases} \frac{4 + \sqrt{100}}{2} = 7 \\ \frac{4 - \sqrt{100}}{2} = -3 \end{cases}$

This equation is also factorable:

$x^2 - 4x - 21 = 0 \rightarrow (x - 7)(x + 3) = 0 \rightarrow \begin{cases} x - 7 = 0 \rightarrow x = 7 \\ x + 3 = 0 \rightarrow x = -3 \end{cases}$

PRACTICES:

Solve each equation by using the quadratic formula.

1) $x^2 + 6x = -8$

2) $3x^2 - 9x - 9 = 3$

3) $6x^2 = 24x - 18$

4) $x^2 = 3x$

5) $3x^2 + 45 = -24x$

6) $2x^2 - 24x = -72$

7) $-6x^2 - 10x - 4 = 8 - 4x^2$

8) $x^2 - 20x = -84$

9) $2x^2 + 18x + 52 = 12$

10) $x^2 + 2x = 15 + 4x$

Score: ....

Answer Key	
1) $\{-4, -2\}$	2) $\{4, -1\}$
3) $\{1, 3\}$	4) $\{3, 0\}$
5) $\{-3, -5\}$	6) $\{6\}$
7) $\{-3, -2\}$	8) $\{14, 6\}$
9) $\{-4, -5\}$	10) $\{5, -3\}$

Chapter 9 : Geometry

Topics that you'll learn in this chapter:

- ➤ The Pythagorean Theorem

- ➤ Area of Triangles and Trapezoids

- ➤ Area and Circumference of Circles

- ➤ Area and Perimeter of Polygons

- ➤ Area of Squares, Rectangles, and Parallelograms

- ➤ Volume of Cubes, Rectangle Prisms, and Cylinder

- ➤ Surface Area of Cubes, Rectangle Prisms, and Cylinder

"Mathematics is, as it were, a sensuous logic, and relates to philosophy as do the arts, music, and plastic art to poetry." — K. Shegel

Name: ....

The Pythagorean Theorem

✓ In any right triangle: $a^2 + b^2 = c^2$

EXAMPLE:

Find the missing length.

Use Pythagorean Theorem: $a^2 + b^2 = c^2$

Then: $a^2 + b^2 = c^2 \rightarrow 3^2 + 4^2 = c^2 \rightarrow 9 + 16 = c^2$

$c^2 = 25 \rightarrow c = 5$

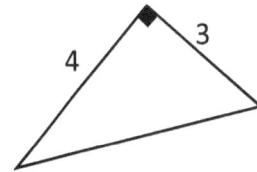

PRACTICES:

Do the following lengths form a right triangle?	Find each missing length to the nearest tenth.
1) 3, 5, 4	2) 32, 60, ?
3) 13, $\sqrt{290}$, 11	4) ?, 72, 35
5) 9, 12, 25	6) 55, 20, ?
7) 5, 13, 12	8) 48, ?, 64

9)	10)
6, 14, 22 (triangle)	?, 13, 5 (right triangle)

Score: ....

Answer Key	
1) Yes	2) 68
3) Yes	4) 62.92
5) No	6) 58.52
7) Yes	8) 80
9) No	10) 12

Name: ....

Angles

- ✓ **Adjacent:** Two triangles are said to be adjacent if its two angles have common side, common vertex and do not overlap.

- ✓ **Vertical:** Two angles share same vertex.

- ✓ **Complementary:** Sum of the measure of two complementary angles are 90°

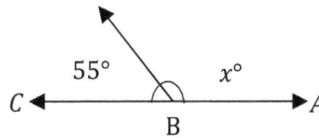

- ✓ **Supplementary:** Sum of the measure of two supplementary angles are 180°

EXAMPLE:

Find x.

Supplementary: Sum of the measure of two supplementary angles are 180°.

Then: $180° - 55° = 125°$

PRACTICES:

What is the value of x in the following figures?

1)

2)

3)

4)

5)	6)
40° x° C←B→A	72° x° C←B→A

7)	8)
x° 160° C←B→A	57° x° C←B→A

Solve.

9) Six supplement peer to each other angles have equal measures. What is the measure of each angle? _____

10) The measure of an angle is one fourth the measure of its complementary. What is the measure of the angle? _____

Score: ....

Answer Key	
1) 60°	2) 91°
3) 32°	4) 25°
5) 50°	6) 18°
7) 20°	8) 123°
9) 30°	10) 18°

Name: ....

Area of Triangles

- ✓ In any triangle the sum of all angles is 180 degrees.
- ✓ Area of a triangle = $\frac{1}{2}$ (base × height)

EXAMPLE:

What is the area of triangle?

Solution:

Use the are formula: Area = $\frac{1}{2}$ (base × height)

base = 12 and height = 8

Area = $\frac{1}{2}(12 \times 8) = \frac{1}{2}(96) = 48$

PRACTICES:

Find the area of each.

1)
c = 15 mi
h = 4 mi

2)
c = 6 m
h = 5.2 m

3)
a = 9.5 m
b = 25 m
c = 18 m
h = 9 m

4)
s = 8 m
h = 6.93 m

5)
c = 25 mi
h = 8 mi

6)
c = 10 m
h = 6.4 m

7) a = 3.5 m b = 7 m c = 16 m h = 7 m	8) s = 12 m h = 4.64 m
9) c = 13 mi h = 6 mi	10) S = 18 m h = 7.4 m

Score: ....

Answer Key

1) 30 mi²	2) 15.6 m²
3) 81 m²	4) 27.72m²
5) 100 mi²	6) 32 m²
7) 56 m²	8) 27.84 m²
9) 39 mi²	10) 133.2 m²

Name: ....

Area of Trapezoids

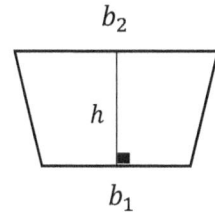

- ✓ A trapezoid is a quadrilateral with at least one pair of parallel sides.
- ✓ Area of a trapezoid = $\frac{1}{2}h(b_1 + b_2)$

EXAMPLE:

Calculate the area of the trapezoid.

Use area formula: $A = \frac{1}{2}h(b_1 + b_2)$

$b_1 = 12$, $b_2 = 16$ and $h = 18$

Then: $A = \frac{1}{2}18(12 + 16) = 9(28) = 252 \; cm^2$

PRACTICES:

Calculate the area for each trapezoid.

1)

2)

3)

4)

5)

6)

7)	8)
5 mi 7 mi 10 mi	7 mm 9.6 mm 4 mm 5 mm

9)	10)
13 cm 10 cm 18 cm	32 m 12 m 40 m

Score: ....

Answer Key

1) 108 cm²	2) 290 m²
3) 350 mi²	4) 71.52 mm²
5) 63 cm²	6) 160 m²
7) 42.5 mi²	8) 24 mm²
9) 155 cm²	10) 432 m²

Name: ....

Area and Perimeter of Polygons

Perimeter of a square
= $4 \times side = 4s$

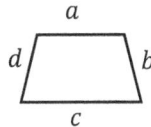 s

Perimeter of a rectangle
= $2(width + length)$

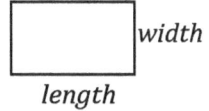 width
length

Perimeter of trapezoid
= $a + b + c + d$

a
d b
c

Perimeter of a regular hexagon = $6a$

 a

Perimeter of a parallelogram = $2(l + w)$

l
w

EXAMPLE:

Find the perimeter of following regular hexagon.

Perimeter of Pentagon = $6a$

Perimeter of Pentagon = $6a = 6 \times 3 = 18m$

3 m
3 m   3 m

PRACTICES:

Find the area and perimeter of each.

1)

15 yd
31.5 yd   31.5 yd
15 yd

2)

11mi
11 mi   11 mi
11 mi

3)

18.4 ft
14.5 ft
12 ft
14.5 ft
18.4 ft

4)

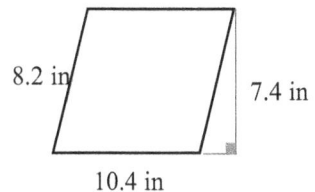

8.2 in   7.4 in
10.4 in

5)

15 cm ◄   ► 10 cm

13 cm

6)

5 mm

8 mm   6 mm

5 mm

Find the perimeter of each shape.

7)

6 m

6 m   6 m

8)

11mm

11 mm

9)

13 ft   13 ft

10)

20 in

19 in

Score: ....

Answer Key

1) Area: 472.5 yd^2, Perimeter: 93 yd	2) Area: 121 mi^2, Perimeter: 44 mi
3) Area: 174 ft^2, Perimeter: 65.8 ft	4) Area: 76.96 in^2, Perimeter: 37.2 in
5) Area: $75cm^2$, Perimeter 52 cm	6) Area: 70 mm^2, Perimeter:38 mm
7) P: 36 m	8) P: 44 mm
9) P: 52 ft	10) P: 78 in

Name: ....

Area and Circumference of Circles

✓ We use variable r for the radius and d for diameter in a circle and π is about 3.14.
✓ Area of a circle= πr^2
✓ Circumference of a circle= $2\pi r$

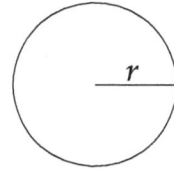

EXAMPLE:

Find the Circumference and area of the circle.

Use Circumference formula: $Circumference = 2\pi r$

$r = 6$, then: $Circumference = 2\pi(6) = 12\pi$

$\pi = 3.14$ then: $Circumference = 12 \times 3.14 = 37.68$

Use area formula: $Area = \pi r^2$,

$r = 6$ then: $Area = \pi(6)^2 = 36\pi$, $\pi = 3.14$ then: $Area = 36 \times 3.14 = 113.04$

PRACTICES:

Find the area and circumference of each. ($\pi = 3.14$)

1)

2)

3)

4)

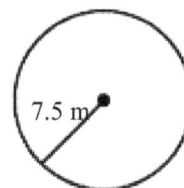

5)

4 m

6)

10 cm

7)

2.5cm

8)

1.5 in

9)

6 km

10)

24 in

Score: ....

Answer Key

1) Area: 12.56 cm^2, Circumference: 12.56cm.	2) Area: 78.5 in2, Circumference: 31.4 in.
3) Area: 200.96 km^2, Circumference: 50.24 km.	4) Area: 176.625 m2, Circumference: 47.1 m.
5) Area: 50.24 m^2, Circumference: 25.12 m	6) Area: 78.5 cm2, Circumference: 31.4 cm.
7) Area: 4.906 cm^2, Circumference: 7.85 cm.	8) Area: 1.766 in2, Circumference: 4.71 in.
9) Area: 113.04 km^2, Circumference:37.68 km.	10) Area: 452.16 in^2, Circumference: 75.36 in

Name: ....

Volume of Cubes

- ✓ A three-dimensional solid object bounded by six square sides is called cube.
- ✓ The measure of the amount of space inside of a solid figure is called volume, like a cube, ball, cylinder, or pyramid.
- ✓ Volume of a cube $= (one\ side)^3$

EXAMPLE:

Find the volume of this cube.

Use volume formula: $volume = (one\ side)^3$

Then: $volume = (one\ side)^3 = (2)^3 = 8\ cm^3$

2 cm

PRACTICES:

Find the volume of each.

1)

2)

3)

4)

5)

6)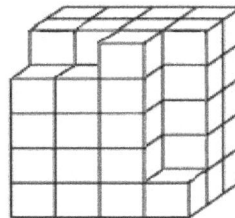

7) 4 ft	8) 6 m
9) 5 in	10) 3 miles

Score: ....

Answer Key

1) 6	2) 34
3) 7	4) 6
5) 41	6) 54
7) $64\,ft^3$	8) $216\,m^3$
9) $125\,in^3$	10) $27\,mi^3$

Name: ....

Volume of Rectangle Prisms

✓ A solid 3-dimensional object which has six rectangular faces.

✓ Volume of a Rectangular prism = **Length × Width × Height**

Volume = $l \times w \times h$

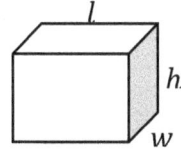

EXAMPLE:

Find the volume and surface area of rectangular prism.

Use volume formula: $Volume = l \times w \times h$

Then: $Volume = 10 \times 5 \times 8 = 400 \; m^3$

PRACTICES:

Find the volume of each of the rectangular prisms.

1)

10 cm
12cm
7 cm

2)

11 cm
9 cm
2 cm

3)

4 m
4 m
4 m

4)
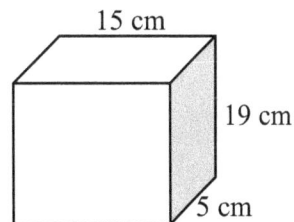
15 cm
19 cm
5 cm

5)

17.5
10 cm
4cm

6)
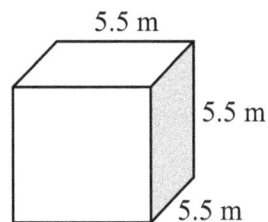
5.5 m
5.5 m
5.5 m

7) 7 m 7 m 7 m	8) 20 ft 11 ft 3 ft
9) 12.5 km 8 km 3 km	10) 8 in 10 in 5 in

Score: ....

Answer Key	
1) 840 cm^3	2) 198 cm^3
3) 64 m^3	4) $1,425 \text{ cm}^3$
5) 700 cm^3	6) 166.375 cm^3
7) 343 cm^3	8) 660 ft^3
9) 300 km^3	10) 400 in^3

Name: ....

Surface Area of Cubes

✓ A three-dimensional solid object bounded by six square sides is called cube.

surface area of cube = **6 × (one side)²**

EXAMPLE:

Find the volume and surface area of this cube.

surface area of cube: $6(one\ side)^2 = 6(2)^2 = 6(4) = 24\ cm^2$

2 cm

PRACTICES:

Find the surface of each cube.

1)

7 mm

2)

10.5 mm

3)

3.5 cm

4)

4 m

5)

3.2 in

6)

8.1 ft

7) 1.6 in	8) 11 m
9) 5.2 in	10) 2.25 mm

Score: ....

Answer Key

1) 294 mm²	2) 661.5 mm²
3) 73.5 cm²	4) 96 m²
5) 61.44 in²	6) 393.66 ft²
7) 15.36 in²	8) 726 m²
9) 162.24 in²	10) 30.375 mm²

Name: ....

Surface Area of a Rectangle Prism

✓ A solid 3-dimensional object which has six rectangular faces.

Surface area= $2(wh + lw + lh)$

EXAMPLE:

Find the volume and surface area of rectangular prism.

Use surface area formula: $Surface\ area = 2(wh + lw + lh)$

Then: $Surface\ area = 2(5 \times 8 + 10 \times 5 + 10 \times 8) =$

$2(40 + 50 + 80) = 340\ m^2$

PRACTICES:

Find the surface of each prism.

1)

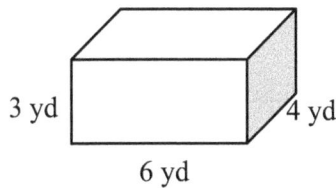

3 yd   4 yd

6 yd

2)

1.02 mm

1.5 mm

0.5 mm

3)

2.5 in

9.5 in

4 in

4)

12 cm

10 cm

7 cm

5)

2 mm

4.5 mm

3.5 mm

6)

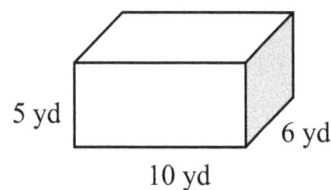

5 yd

6 yd

10 yd

| 7) 14 cm, 8 cm, 5 cm | 8) 2 yd, 6 yd, 5 yd |
| 9) 3.5 in, 7.5 in, 6 in | 10) 4 mm, 8 mm, 6 mm |

Score: ....

Answer Key

1) 108 yd²	2) 5.58 mm²
3) 143.5 in²	4) 548 cm²
5) 63.5 mm²	6) 280 yd²
7) 444 cm²	8) 104 yd²
9) 184.5 in²	10) 208 mm²

Name: ....

Volume of a Cylinder

- ✓ A solid geometric figure with straight parallel sides and a circular cross section is called a cylinder.
- ✓ Volume of Cylinder Formula = $\pi(radius)^2 \times height$

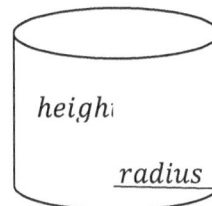

  $\pi = 3.14$

height

radius

EXAMPLE:

Find the volume of the follow Cylinder.

Use volume formula: $Volume = \pi(radius)^2 \times height$

Then: $Volume = \pi(4)^2 \times 6 = \pi16 \times 6 = 96\pi$

$\pi = 3.14$ then: $Volume = 96\pi = 301.44$

6 cm

4 cm

PRACTICES:

Find the volume of each cylinder. ($\pi = 3.14$).

1)

4 in

6 in

2)

7 m

10 m

3)

3 m

6 m

4)

2 in

4.5 in

5)

7.5 m

4 m

6)

14 in

3.5 in

7) 10 in, 7.5 in	8) 6 ft, 10 ft
9) 5 in, 9 in	10) 12 yd, 2 yd

Score: ....

Answer Key	
1) 301.44 in³	2) 1538.6 m³
3) 42.39 m³	4) 14.13 in³
5) 376.8 m³	6) 538.51 in³
7) 588.75 in³	8) 282.6 ft³
9) 706.5 in³	10) 150.72 yd³

Name: ....

Surface Area of a Cylinder

✓ A solid geometric figure with straight parallel sides and a circular cross section is called cylinder.

✓ Surface area of a cylinder= $2\pi r^2 + 2\pi rh$

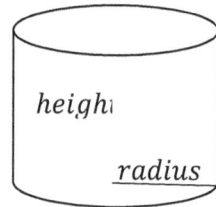

height

radius

EXAMPLE:

Find the Surface area of the follow Cylinder.

Use surface area formula: $Surface\ area = 2\pi r^2 + 2\pi rh$

Then: $= 2\pi(4)^2 + 2\pi(4)(6) = 2\pi(16) + 2\pi(24) = 32\pi + 48\pi =$

80π

$\pi = 3.14$ then: $Surface\ area = 80 \times 3.14 = 251.2$

6 cm

4 cm

PRACTICES:

Find the surface of each cylinder. ($\pi = 3.14$).

1)	2)
5 ft 8 ft	7 cm 4cm
3) 6 in 10 in	4) 2 yd 5.5 yd
5) 18 in 12 in	6) 1.5 m 4 m

7)	8)
9)	10)

Score: ....

Answer Key

1) 226.08 ft²	2) 113.04 cm²
3) 224.92 in²	4) 94.2 yd²
5) 1,186.92 in²	6) 51.81 m²
7) 244.92 in²	8) 502.4 ft²
9) 70.336 yd²	10) 785 in²

Chapter 10 : Statistics

Topics that you'll learn in this chapter:

- ➢ Mean, Median, Mode, and Range of the Given Data

- ➢ Box and Whisker Plots

- ➢ Bar Graph

- ➢ Stem– And– Leaf Plot

- ➢ The Pie Graph or Circle Graph

- ➢ Dot and Scatter Plots

- ➢ Probability of Simple Events

"The book of nature is written in the language of Mathematic" -Galileo

Name: ....

Mean and Median

- ✓ Mean: $\dfrac{\text{sum of the data}}{\text{total number of data entires}}$
- ✓ Median: Middle value in the sorted list of numbers
- ✓ When there are two middle numbers, we average them

EXAMPLE:

What is the median of these numbers? $4, 9, 13, 8, 15, 18, 5, 11$

Write the numbers in order: $4, 5, 8, 9, 11, 13, 15, 18$

Median is the number in the middle. Therefore, there are 9 and 11 in the middle, then

find the average: $\dfrac{9+11}{2} = \dfrac{20}{2} = 10$, the median is 10

PRACTICES:

Find Mean and Median of the Given Data.

1) $8, 10, 7, 3, 12$	2) $4, 6, 9, 7, 5, 19$
3) $5, 11, 1, 1, 8, 9, 20$	4) $12, 4, 2, 7, 3, 2$
5) $3, 5, 7, 4, 7, 8, 9$	6) $5, 10, 4, 4, 9, 12, 9$
7) $10, 4, 8, 5, 9, 6, 7, 19$	8) $16, 3, 4, 3, 7, 6, 18$

Solve.

9) In a javelin throw competition, five athletics score 23, 45,53.53,13and 61 meters. What are their Mean and Median? _____

10) Eva went to shop and bought 7 apples, 4 peaches, 6 bananas, 3 pineapples and 4melons. What are the Mean and Median of her purchase? _____

Score: ....

Answer Key	
1) Mean: 8, Median: 8	2) Mean: 8.33, Median: 6.5
3) Mean: 7.85, Median: 8	4) Mean: 5, Median: 3.5
5) Mean: 6.14, Median: 7	6) Mean: 7.57, Median: 9
7) Mean: 8.5, Median: 7.5	8) Mean: 8.14, Median: 6
9) Mean: 39.106, Median: 45	10) Mean: 4.8, Median: 4

Name: ....

Mode and Range

✓ Mode: The most appeared value in the list.

✓ Range: The difference of highest value and lowest value in the list

EXAMPLE:

What is the mode(s) of these numbers? **22, 16, 12, 9, 7, 6, 4, 6, 9**

Mode: The most appeared value in the list.

Therefore: modes are 6 and 9

PRACTICES:

Find Mode and Rage of the Given Data.

1) 10, 12, 8, 8,4, 1, 9 Mode: _____Range: _____	2) 4, 6, 4, 13, 2, 13, 19, 13 Mode: _____Range: _____
3) 8, 8, 7, 2, 7, 7, 5, 6, 5 Mode: _____Range: _____	4) 12, 9, 12,6, 12, 9, 10 Mode: _____Range: _____
5) 2, 2, 4, 3, 2, 10, 8 Mode: _____Range: _____	6) 6, 1, 4, 20, 19, 2, 7, 1, 5, 1 Mode: _____Range: _____
7) 16, 35, 9, 7, 7, 5, 14, 13, 7 Mode: _____Range: _____	8) 7, 6, 6, 9, 16, 6, 7, 5 Mode: _____Range: _____

Solve.

9) A stationery sold 15 pencils, 26 red pens, 22 blue pens, 10 notebooks, 12 erasers, 22 rulers and 42 color pencils. What are the Mode and Range for the stationery sells?

Mode: _____ Range: _____

10) In an English test, eight students score 24, 13, 17, 21, 19, 13, 13 and 17. What are their Mode and Range? _____

Score: ....

Answer Key	
1) Mode: 8, Range: 11	2) Mode: 13, Range: 17
3) Mode: 7, Range: 6	4) Mode: 12, Range: 6
5) Mode: 2, Range: 8	6) Mode: 1, Range: 19
7) Mode: 7, Range: 30	8) Mode: 6, Range: 11
9) Mode: 22, Range: 32	10) Mode: 13, Range: 11

Name: ....

Times Series

✓ A precise representation of the distribution of numerical data is referred as Time Series.

EXAMPLE:

Use the following Graph to complete the table.

Day	Distance (km)
1	
2	

→

Answer:

Day	Distance (km)
1	359
2	460
3	278
4	547
5	360

PRACTICES:

Use the following Graph to complete the table.

1)

Day	Distance (km)
1	
2	

The following table shows the number of births in the US from 2007 to 2012 (in millions).

2)

Draw a time series for the table.

Year	Number of births (in millions)
2007	6.42
2008	6.45
2009	6.33
2010	5.9
2011	4.35
2012	4.35

Score: ....

Answer Key

1)

Day	Distance (km)
1	343
2	430
3	268
4	507
5	390

2)

Number of Births

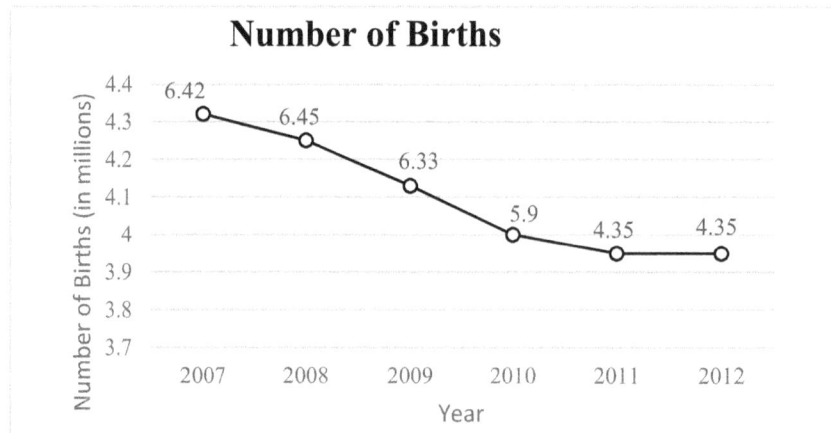

Name: ....

Box and Whisker Plot

- ✓ Box-and-whisker plots display data including quartiles.
- ✓ IQR – interquartile range shows the difference from Q1 to Q3.
- ✓ Extreme Values are the lowest and highest values in a data set.

EXAMPLE:

$73, 84, 86, 95, 68, 67, 100, 94, 77, 80, 62, 79$

Maximum: 100, Minimum: 62; Q_1: 70.5; Q_2: 79.5; Q_3: 90

PRACTICES:

Make box and whisker plots for the given data.

1) $1, 5, 20, 8, 3, 10, 13, 11, 14, 17, 18, 15, 23$

2) $2, 7, 23, 11, 13, 9, 16, 5, 18, 22, 20, 17, 19$

3) $3, 7, 9, 10, 11, 5, 14, 19, 20, 21, 22, 8, 14$

4) $4, 6, 5, 15, 12, 14, 10, 7, 21, 17, 8, 22, 6$

Score: ....

Answer Key

1) 1,3, 5, 8, 10, 11, 13, 14, 15, 17,18, 20, 23

Maximum: 23, Minimum: 1, Q_1: 8, Q_2: 13, Q_3: 17

2) 2, 7, 23, 11, 13, 9, 16, 5, 18, 22, 20, 17, 19

Maximum: 23, Minimum: 2, Q_1: 9, Q_2: 16, Q_3: 19

3) 3,7, 9, 10, 11, 5, 14, 19, 20, 21 ,22, 8, 14

Maximum: 22, Minimum: 2, Q_1: 8, Q_2: 11, Q_3: 19

4) 4, 6, 5, 15, 12, 14, 10, 7, 21, 17, 8, 22, 6

Maximum: 22, Minimum: 4, Q_1: 6, Q_2: 10, Q_3: 15

Name: ....

Bar Graph

✓ A chart that presents data with bars in different heights to match with the values of the data is called a bar graph. We can graph the bars vertically or horizontally.

EXAMPLE:

Graph the given information as a bar graph.

Name of the Sport	Total Number of Students
Football	15
Volleyball	7
Table Tennis	7
Basketball	12
Chess	9

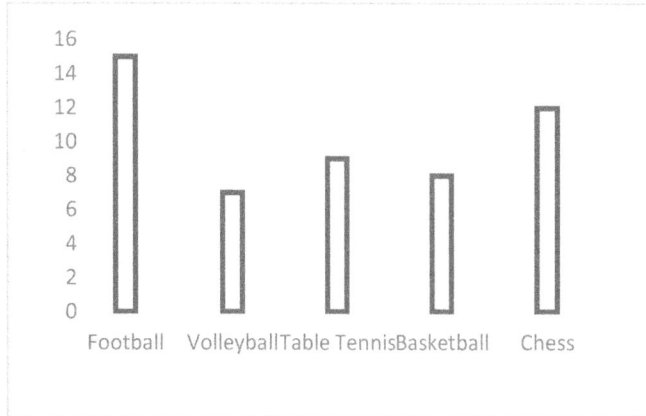

PRACTICES:

Graph the given information as a bar graph.

1)

Day	Sale House
Monday	6
Tuesday	4
Wednesday	10
Thursday	5
Friday	2
Saturday	8
Sunday	1

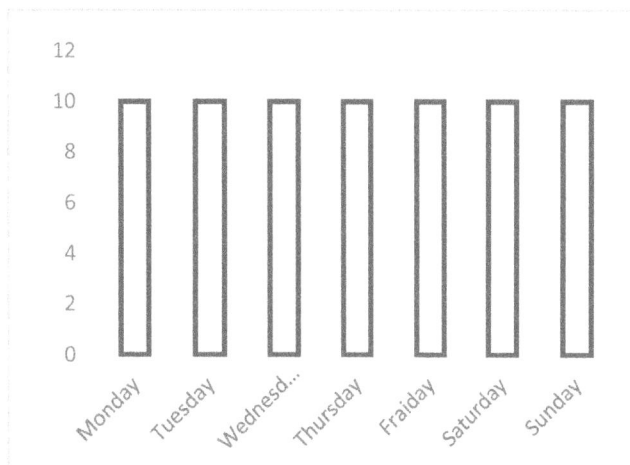

2)

Day	Sale House
Monday	8
Tuesday	6
Thursday	3
Friday	10
Saturday	4
Sunday	2

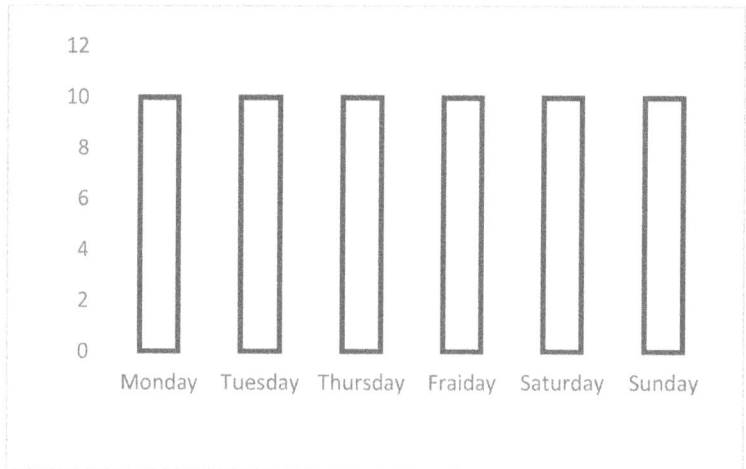

Score: ....

Answer Key

1)

2)

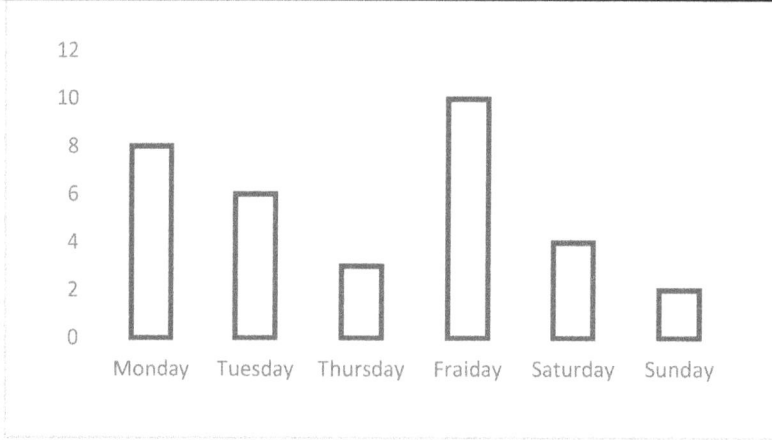

Dot plots

✓ The representation of a distribution that consists of group of data points plotted on a simple scale is referred as a dot plot. Dot plots are used for continuous, univariate and quantitative data. If there are few data points, then they can be labelled.

✓ Dot plots are one of the simplest statistical plots and are suitable for small to moderate sized data sets.

EXAMPLE:

A survey of "How many books each student purchased?" has these results How many students purchase 4 books?

4 students

PRACTICES:

A survey of "How many pets each person owned?" has these results:

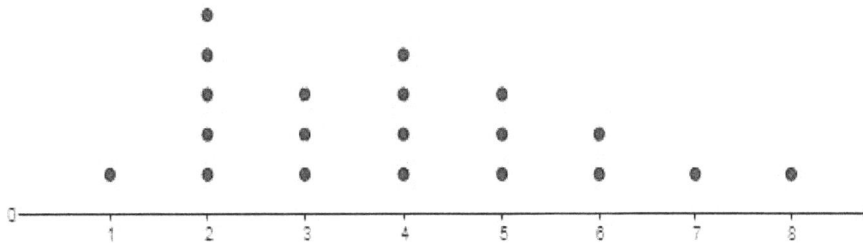

1) How many people have at least 3 pets?

2) How many people have 2 and 3 pets?

3) How many people have 4 pets?

4) How many people have 2 or less than 2 pets?

5) How many people have more than 7 pets?

6) How many people have more than 4 pets?

Score: ....

Answer Key	
1) 4	2) 8
3) 4	4) 6
5) 1	6) 7

Name: ....

Scatter Plots

- ✓ The values with points that represent the relationship between two sets of data are shown by a scatter(x, y) plot.
- ✓ The horizontal values are taken as x and vertical data is taken as y.

EXAMPLE:

Construct a scatter plot.

x	1	2	3	4	5
y	2	4.5	1.5	5	2

PRACTICES:

Construct a scatter plot.

1)

x	1	2.5	3	3.5	4	5
y	4	3.5	4.5	2.5	8	2

2)

x	1	2	3.5	4	4.5	5
y	3	1	2.5	1.5	1.5	1

Y-Values

Score: ....

Answer Key

1)

Y-Values

2)

Y-Values

Name: ....

Stem-And-Leaf Plot

- ✓ Stem-and-leaf plots display the frequency of the values in a data set.
- ✓ We can make a frequency distribution table for the values, or we can use a stem-and-leaf plot.

EXAMPLE:

56, 58, 42, 48, 66, 64, 53, 69, 45, 72

Stem	leaf
4	2 5 8
5	3 6 8
6	4 6 9
7	2

PRACTICES:

Make stem ad leaf plots for the given data.

1) 42, 14, 17, 21, 44, 24, 18, 47, 23, 24, 19, 12

Stem	Leaf plot

2) 10, 65, 14, 18, 69, 11, 33, 61, 66, 38, 15, 35

Stem	Leaf plot

3) 122, 87, 99, 86, 100, 126, 92, 129, 88, 121, 91, 107

Stem	Leaf plot

4) 60, 51, 119, 69, 72, 59, 110, 65, 77, 59, 65, 112, 71

Stem	Leaf plot

Score: ....

Answer Key

1)

Stem	leaf
1	2 4 7 8 9
2	1 3 4
4	2 4 7

2)

Stem	leaf
1	0 1 4 5 8
3	3 5 8
6	1 5 6 9

3)

Stem	leaf
8	6 7 8
9	1 2 9
10	0 7
12	1 2 6 9

4)

Stem	leaf
5	1 9 9
6	0 5 5 9
7	1 2 7
11	0 2 9

Name: ....

The Pie Graph or Circle Graph

✓ A circular chart divided into sectors is said to be a pie chart; each sector represents the relative size of each value.

EXAMPLE:

A library has 840 books that include Mathematics, Physics, Chemistry, English, and

History. Use following graph to answer question.

What is the number of Mathematics books?

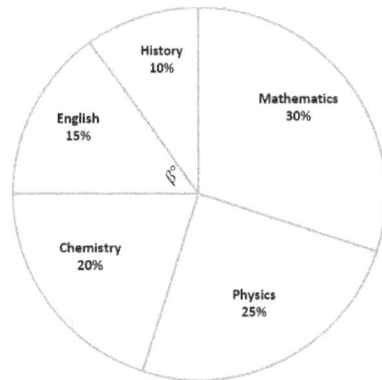

Number of total books = 840,

Percent of Mathematics books = 30% = 0.30

Then: $0.30 \times 840 = 252$

PRACTICES:

Favorite Sports:

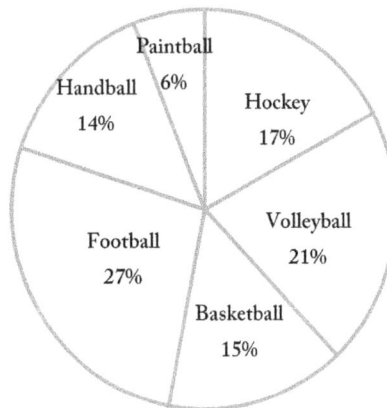

SPORTS

1) What percentage of pie graph is paintball?

2) What percentage of pie graph is Hockey and volleyball?

3) What percentage of pie not Football and Basketball?

4) What percentage of pie graph is Hockey and Handball and Football?

5) What percentage of pie graph is Basketball?

6) What percentage of pie not Handball and Paintball?

Score: ....

Answer Key

1) 6%	2) 38%
3) 58%	4) 58%
5) 15%	6) 80%

Name: ....

Probability of Simple Events

✓ Probability is the possibility of something happening in the future. It is shown as a number between zero (can never happen) to 1 (will always happen).

✓ Probability can be written as a fraction, a decimal, or a percent.

EXAMPLE:

If there are 8 red balls and 12 blue balls in a basket, what is the probability that John will pick out a red ball from the basket?

There are 8 red ball and 20 are total number of balls. Therefore, probability that John will pick out a red ball from the basket is 8 out of 20 or $\frac{8}{8+12} = \frac{8}{20} = \frac{2}{5}$

PRACTICES:

Solve.

1) A number is chosen at random from 28 to 35. Find the probability of selecting factors of 5.

2) A number is chosen at random from 1 to 60. Find the probability of selecting multiples of 15.

3) Find the probability of selecting 4queens from a deck of card.

4) A number is chosen at random from 8 to 19. Find the probability of selecting factors of 3.

5) What probability of selecting a ball less than 6 from 10 different bingo balls?

6) A number is chosen at random from 1 to 10. What is the probability of selecting a multiple of 2?

7) A card is chosen from a well-shuffled deck of 52 cards. What is the probability that the card will be a king OR a queen?

8) A number is chosen at random from 1 to 20. What is the probability of selecting multiples of 5?

9) A number is chosen at random from 1 to 10. Find the probability of selecting number 4 or smaller numbers.

10) Bag A contains 9 red marbles and 3 green marbles. Bag B contains 9 black marbles and 6 orange marbles. What is the probability of selecting a green marble at random from bag A? What is the probability of selecting a black marble at random from Bag B?

Score: ....

Answer Key

1) $\frac{1}{4}$	2) $\frac{1}{15}$
3) $\frac{1}{13}$	4) $\frac{1}{3}$
5) $\frac{1}{2}$	6) $\frac{1}{2}$
7) $\frac{2}{13}$	8) $\frac{1}{5}$
9) $\frac{2}{5}$	10) $\frac{1}{4}, \frac{3}{5}$

Chapter 11 : GED Test Review

The General Educational Development Test, commonly known as the GED or high school equivalency degree, is a standardized test and is the only high school equivalency test recognized in all 50 USA states.

Currently, GED is a computer-based test. Official computer-based tests are given at test centers all over the country.

There are four subject area tests on GED:

- o Reasoning Through Language Arts,
- o Mathematical Reasoning,
- o Social Studies,
- o Science

The GED Mathematical Reasoning test is a 115-minute, single-section test that covers basic mathematics topics, quantitative problem-solving and algebraic questions. There are two parts on Mathematical Reasoning section. The first part contains 5 questions where calculators are not permitted. The second part contains 41 test questions. Calculator is allowed in the second part.

In this section, there are two complete GED Mathematical Reasoning Tests. Take these tests to see what score you'll be able to receive on a real GED test.

Time to Test

Time to refine your skill with a practice examination

Take a REAL GED Mathematical Reasoning test to simulate the test day experience. After you have finished, score your test using the answer key.

Before You Start

- You will need a pencil, calculator, and a timer to take the test.

- It is okay to guess. You won't lose any points if you're wrong.

- After you have finished the test, review the answer key to see where you went wrong.

Calculators are permitted only for the Part 2 of the test.

Good Luck!

GED Test Mathematics Formula Sheet

Area of a:	
Parallelogram	$A = bh$
Trapezoid	$A = \frac{1}{2}h(b_1 + b_2)$

Surface Area and Volume of a:		
Rectangular/Right Prism	$SA = ph + 2B$	$V = Bh$
Cylinder	$SA = 2\pi rh + 2\pi r^2$	$V = \pi r^2 h$
Pyramid	$SA = \frac{1}{2}ps + B$	$V = \frac{1}{3}Bh$
Cone	$SA = \pi rs + \pi r^2$	$V = \frac{1}{3}\pi r^2 h$
Sphere	$SA = 4\pi r^2$	$V = \frac{4}{3}\pi r^3$
		(p = perimeter of base B; $\pi = 3.14$)

Algebra	
Slope of a line	$m = \frac{y_2 - y_1}{x_2 - x_1}$
Slope-intercept form of the equation of a line	$y = mx + b$
Point-slope form of the Equation of a line	$y - y_1 = m(x - x_1)$
Standard form of a Quadratic equation	$y = ax^2 + bx + c$
Quadratic formula	$x = \frac{-b \pm \sqrt{b^2 - 4ac}}{2a}$
Pythagorean theorem	$a^2 + b^2 = c^2$
Simple interest	$I = prt$ (I= interest, p= principal, r= rate, t= time)

GED Practice Test 1

Mathematical Reasoning

Two Parts

Total number of questions: 46

Part 1 (Non-Calculator): 5 questions

Part 2 (Calculator): 41 questions

Total time for two Part: 115 Minutes

Administered *Month Year*

GED Practice Test 1

Mathematical Reasoning

Part 1: Non-Calculator

5 Questions

You may NOT use a calculator on this part.

Administered *Month Year*

1) A tree 45 feet tall casts a shadow 28 feet long. Jack is 9 feet tall. How long is Jack's shadow?

A. 5.6 ft

B. 4.5 ft

C. 6.5 ft

D. 6.8 ft

2) What is the value of the expression $-3(x + 2y) + 2(2 - x)^2$ when $x = 1$ and $y = -3$?

A. -17

B. -15

C. 17

D. 15

3) What is the slope of a line that is perpendicular to the line $2x - 3y = 6$?

A. -3

B. $-\frac{3}{2}$

C. $\frac{3}{2}$

D. 3

4) $[5 \times (-14)] + ([(-6) \times (-11)] \div 6) + (-17) = ?$

Write your answer in the box below.

5) What is the product of all possible values of x in the following equation?

$$|-6x + 3| = 33$$

A. -5

B. -30

C. 30

D. 7

GED Practice Test 1

Mathematical Reasoning

Part 2: Calculator

41 Questions

You may use a calculator on this part.

Administered *Month Year*

6) Simplify the expression.

$$(2x^4 - x^2 - x^3) - (2x^2 - 3x^4 + x^3)$$

A. $-(3x^4 + 3x^3 - 3x^2)$

C. $3(x^4 + 3x^3 - 3x^2)$

B. $(5x^4 - 2x^3 - 3x^2)$

D. $(-5x^4 - 5x^3 - 3x^2)$

7) What is the value of 3^6?

Write your answer in the box below.

8) Last week 30,000 fans attended a football match. This week Two times as many bought tickets, but one sixth of them cancelled their tickets. How many are attending this week?

A. 60,000

C. 50,000

B. 80,000

D. 75,000

9) In two successive years, the population of a town is increased by 12% and 18%. What percent of the population is increased after two years?

A. 16.32%

C. 32.2%

B. 34.2%

D. 20 %

10) Which of the following graphs represents the compound inequality $8 \leq 2x + 6 < 18$?

A.

B.

C.

D.

11) Two dice are thrown simultaneously, what is the probability of getting a sum of 3 or 5 or 7?

A. $\frac{1}{2}$ C. $\frac{4}{36}$

B. $\frac{1}{3}$ D. $\frac{5}{18}$

12) A swimming pool holds 3,276 cubic feet of water. The swimming pool is 36 feet long and 13 feet wide. How deep is the swimming pool?

Write your answer in the box below. (Don't write the measurement)

13) The mean of 25 test scores was calculated as 85. But it turned out that one of the scores was misread as 45 but it was 20. What is the correct mean of the test scores?

A. 84

C. 85.5

B. 85

D. 84.5

14) Which of the following shows the numbers in descending order?

$$\frac{1}{12}, 0.5, 45\%, \frac{1}{5}$$

A. $\frac{1}{12}, 0.5, \frac{1}{5}, 45\%$

C. $0.5, 45\%, \frac{1}{5}, \frac{1}{12}$

B. $45, \% \frac{1}{12}, \frac{1}{5}, 0.5$

D. $\frac{1}{12}, \frac{1}{5}, 45\%, 0.5$

15) What is the volume of a box with the following dimensions?

Hight = 5 cm   Width = 8 cm   Length = 3 cm

A. 100 cm³

C. 60 cm³

B. 120 cm³

D. 40 cm³

16) The average of 6 numbers is 28. The average of 4 of those numbers is 35. What is the average of the other two numbers?

A. 15

C. 24

B. 14

D. 25

17) What is the value of x in the following system of equations?

$$4x + y = -2$$
$$-x - 2y = -3$$

A. 3   C. 4

B. 2   D. -3

18) The perimeter of a rectangular yard is 112 meters. What is its length if its width is triple its length?

A. 20 meters   C. 14 meters

B. 16 meters   D. 18 meters

19) In a stadium the ratio of home fans to visiting fans in a crowd is 5:7. Which of the following could be the total number of fans in the stadium? (Select one or more answer choices)

A. 49,200   D. 51,400

B. 44,600   E. 63,000

C. 54,400

20) What is the perimeter of a square in centimeters that has an area of 256.32 cm^2?

Write your answer in the box below. (don't write the measurement)

[ ]

21) Anita's trick–or–treat bag contains 18 pieces of chocolate, 21 suckers, 11 pieces of gum, 12 pieces of licorice. If she randomly pulls a piece of candy from her bag, what is the probability of her pulling out a piece of gum?

A. $\frac{6}{25}$

C. $\frac{11}{62}$

B. $\frac{1}{13}$

D. $\frac{12}{25}$

22) What is the area of a square whose diagonal is 8?

A. 64

C. 34

B. 18

D. 32

23) Which of the following points lies on the line $4x - 2y = -4$? (Select one or more answer choices)

A. $(-1, 4)$

C. $(-2, 2)$

B. $(1, 4)$

D. $(0, 2)$

24) Mr. Carlos family are choosing a menu for their reception. They have 3choices of appetizers, 5 choices of entrees, 8 choices of cake. How many different menu combinations are possible for them to choose?

A. 120

C. 95

B. 102

D. 210

25) The ratio of boys and girls in a class is 9:11. If there are 100 students in the class, how many more boys should be enrolled to make the ratio 1:1?

A. 11

B. 9

C. 10

D. 5

26) The average of six numbers is 42. If a seventh number that is greater than 54 is added, then, which of the following could be the new average?

A. 41

B. 43

C. 44

D. 42

27) The perimeter of the trapezoid below is 55 cm. What is its area?

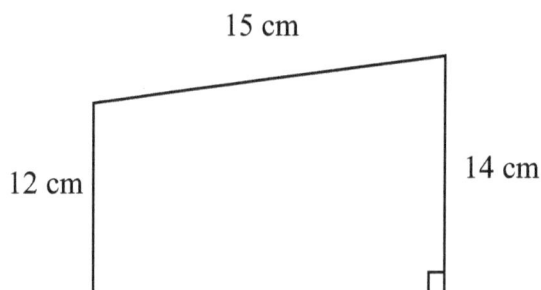

15 cm

12 cm

14 cm

A. 192 cm^2

B. 164 cm^2

C. 182 cm^2

D. 168 cm^2

28) A card is drawn at random from a standard 52–card deck, what is the probability that the card is of Diamonds and Hearts? (The deck includes 13 of each suit clubs, diamonds, hearts, and spades).

A. $\frac{1}{26}$

B. $\frac{1}{4}$

C. $\frac{1}{2}$

D. $\frac{2}{13}$

29) The diagonal of a rectangle is 9 inches long and the height of the rectangle is 4 inches. What is the perimeter of the rectangle in inches?

Write your answer in the box below.

30) What is the surface area of the cylinder below?

A. 288 π in²

B. 352.14 π in²

C. 264 π in²

D. 344 π in²

12 m

18 m

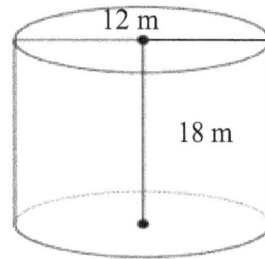

31) Simplify $4x^3y^4(-xy^3)^3 =$

A. $4x^6y^{10}$

C. $8x^7y^{12}$

B. $-4x^7y^{12}$

D. $-4x^6y^{13}$

32) Mr. Matthews saves $3,200 out of his annually family income of $60,800. What fractional part of his income does he save?

A. $\frac{2}{17}$

C. $\frac{5}{19}$

B. $\frac{1}{19}$

D. $\frac{2}{3}$

33) What is the median of these numbers? 24,1 8,13, 11, 14, 6, 5

    A. 5.2   C. 13

    B. 9.5   D. 8

34) A bank is offering 3.15% simple interest on a savings account. If you deposit $21,000, how much interest will you earn in six years?

    A. $3400   C. $3,969

    B. $11,640   D. $26,400

35) Mrs. Thomson needs an 90% average in her writing class to pass. On her first 3 exams, he earned scores of 87%, 82%, and 98%. What is the minimum score Mrs. Thomson can earn on her fourth and final test to pass?

Write your answer in the box below.

☐

36) Daniel is 10 miles ahead of Noa and running at 2 miles per hour. Noa is running at the speed of 4 miles per hour. How long does it take Noa to catch Daniel?

    A. 1 hour, and 40 minutes   C. 6 hours

    B. 2 hours, 30 minutes   D. 5 hours

37) What is the equivalent temperature of $93°F$ in Celsius?

$$C = \frac{5}{9}(F - 32)$$

A. 43

C. 46.66

B. 33.88

D. 27.4

38) A football team had $31,000 to spend on supplies. The team spent $15,800 on new balls. New sport shoes cost $95 each. Which of the following inequalities represent the number of new shoes the team can purchase?

A. $95x + 15,800 \leq 31,000$

C. $15,000 + 95x \geq 31,000$

B. $95x + 18,500 \leq 31,000$

D. $15,800 + 95x \geq 31,000$

39) If 70 % of a number is 36, what is the number?

A. 51.43

C. 25

B. 250

D. 50

40) The circle graph below shows all Mr. Wilson's expenses for last month. If he spent $720 on his car, how much did he spend for his rent?

A. $1,18.81

Mr. Green's monthly expenses

B. $818.18

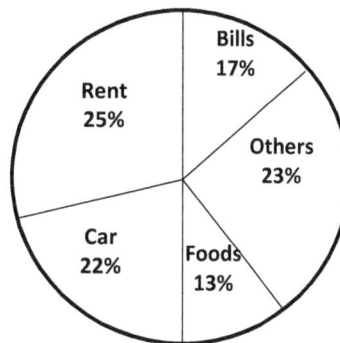

C. $1880

D. $1,820

41) 175 students took an exam and 65 of them failed. What percent of the students passed the exam?

A. 50 %

C. 66.78 %

B. 88.67 %

D. 62.86 %

42) A cruise line ship left Port A and traveled 30 miles due west and then 40 miles due north. At this point, what is the shortest distance from the cruise to port A in miles?

Write your answer in the box below.

43) The square of a number is $\frac{80}{125}$. What is the cube of that number?

A. $\frac{16}{25}$

C. $\frac{8}{25}$

B. $\frac{80}{125}$

D. $\frac{64}{125}$

44) What is the value of x in the following equation?

$$\frac{1}{3}x + \frac{3}{7} = \frac{1}{2}$$

A. 5

C. $\frac{3}{14}$

B. $\frac{3}{7}$

D. $\frac{5}{14}$

45) If 120 % of a number is 85, then what is the 60 % of that number?

A. 42.49

C. 50.6

B. 47.5

D. 61.3

46) What is the slope of the line?

$$-4x - 5y = 32$$

Write your answer in the box below.

[   ]

End of GED Mathematical Reasoning Practice Test 1

GED Practice Test 2

Mathematical Reasoning

Two Parts

Total number of questions: 46

Part 1 (Non-Calculator): 5 questions

Part 2 (Calculator): 41 questions

Total time for two Part: 115 Minutes

Administered *Month Year*

GED Practice Test 2

Mathematical Reasoning

Part 1: Non-Calculator

5 Questions

You may NOT use a calculator on this part.

Administered *Month Year*

1) Which of the following points lies on the line with equation $x - \frac{3}{2}y = -4$?

    A. $(-1, -2)$ C. $(-2, 1)$

    B. $(-1, 2)$ D. $(1, -2)$

2) A shirt costing \$250 is discounted 15%. After a month, the shirt is discounted another 5%. Which of the following expressions can be used to find the selling price of the shirt?

    A. $(250)(0.85)(0.95)$ C. $(250)(0.85) - (240)(0.10)$

    B. $(250)(0.15)(0.05)$ D. $(250)(0.80) - (250)(0.90)$

3) If $-2x + 9 = 1.5$, What is the value of $3x - \frac{3}{2}$?

    A. 9.5 C. -9.75

    B. -9.5 D. 9.75

4) $4 \times (-3) + 1 + 2(-8 - 9 \times 3) \div 7 = ?$

Write your answer in the box below.

```
┌─────────────────────────┐
│                         │
└─────────────────────────┘
```

5) What is the area of an isosceles right triangle that has one leg that measures 6cm?

    A. 21 cm^2 C. 24 cm^2

    B. 18 cm^2 D. 32 cm^2

GED Practice Test 2

Mathematical Reasoning

Part 2: Calculator

41 Questions

You may use a calculator on this part.

Administered *Month Year*

6) Ryan traveled 250 km in 5 hours and Riley traveled 180 km in 3 hours. What is the ratio of the average speed of Ryan to average speed of Riley?

A. 8: 5   C. 5: 6

B. 6: 5   D. 8: 6

7) What is the value of y in the following system of equation?

$$3x + 4y = -9$$

$$-5x - 2y = -6$$

Write your answer in the box below.

□

8) An angle is equal to one sixth of its supplement. What is the measure of that angle?

A. 15   C. 25.71

B. 18.71   D. 24

9) Abigail purchased a sofa for $490.5. The sofa is regularly priced at $552. What was the percent discount Abigail received on the sofa?

A. 11.14 %   C. 20 %

B. 22.14 %   D. 12 %

www.mathnotion.com 277

10) When a number is subtracted from 40 and the difference is divided by that number, the result is 3. What is the value of the number?

A. 10   C. 8

B. 17   D. 9

11) Right triangle ABC has two legs of lengths 9 cm (AB) and 12 cm (AC). What is the length of the third side (BC)?

A. 15 cm   C. 30 cm

B. 16 cm   D. 22 cm

12) A taxi driver earns \$18 per hour work. If he works 10 hours a day, and he uses 3-liters Petrol in 2 hours with price \$3.50 for 1-liter. How much money does he earn in one day?

A. \$135.5   C. \$120

B. \$187   D. \$127.5

13) The width of a box is one fourth of its length. The height of the box is one half of its width. If the length of the box is 40 cm, what is the volume of the box?

A. 2560 cm^3 C. 2000 cm^3

B. 2768 cm^3 D. 2024 c m^3

14) The price of a sofa is decreased by 30% to $770. What was its original price?

    A. $1,200   C. $1,600

    B. $1,640   D. $1,100

15) If 70 % of a class are girls, and 40 % of girls play tennis, what percent of the class play tennis?

    A. 14 %   C. 21 %

    B. 28 %   D. 39 %

16) The average of 29, 27, 20 and x is 25. What is the value of x?

Write your answer in the box below.

[ ]

17) The price of a car was $32,000 in 2014, $38,400 in 2015 and $20,480 in 2016. What is the rate of depreciation of the price of car per year?

    A. 20 %   C. 25 %

    B. 35 %   D. 15 %

18) The average of three consecutive numbers is 45. What is the smallest number?

    A. 45   C. 44

    B. 43   D. 42

19) The area of a circle is less than 64 π. Which of the following can be the circumference of the circle? (Select one or more answer choices)

    A. 14 π D. 20 π

    B. 19 π E. 12 π

    C. 18 π

20) A bank is offering 2.5% simple interest on a savings account. If you deposit $7500, how much interest will you earn in two years?

    A. $375   C. $187

    B. $187.5   D. $3750

21) In four successive hours, a car travels 58 km, 54 km, 52 km, and 50km. In the next four hours, it travels with an average speed of 55 km per hour. Find the total distance the car traveled in 8 hours.

    A. 346 km   C. 432 km

    B. 434 km   D. 444 km

22) The ratio of boys to girls in a school is 5:8. If there are 338 students in a school, how many boys are in the school.

Write your answer in the box below.

[ ]

23) How long does a 245–miles trip take moving at 25 miles per hour (mph)?

A. 9 hours

C. 9 hours and 48 minutes

B. 9 hours and 36 minutes

D. 7 hours and 48 minutes

24) The perimeter of the trapezoid below is 60. What is its area?

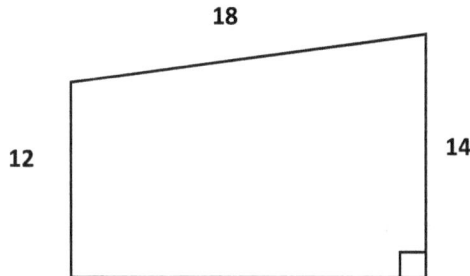

18

12

14

Write your answer in the box below.

25) In the xy-plane, the point $(1, -1)$ and $(2, 1)$ are on the line A. Which of the following points could also be on the line A? (Select one or more answer choices)

A. $(1, 5)$

D. $(0, -4)$

B. $(2, 5)$

E. $(-2, -7)$

C. $(0, -3)$

26) 54 is What percent of 48?

A. 122 %

C. 121.5 %

B. 112.5 %

D. 136 %

27) The marked price of a computer is E Euro. Its price decreased by 20% in March and later increased by 5 % in April. What is the final price of the computer in E Euro?

A. 1.85 E

C. 0.85 E

B. 0.84 E

D. 1.05 E

28) A rope weighs 350 grams per meter of length. What is the weight in kilograms of 15 meters of this rope? (1 kilograms = 1000 grams)

A. 5.25

C. 52.5

B. 15.25

D. 6.25

29) Which of the following could be the product of two consecutive prime numbers? (Select one or more answer choices)

A. 77

D. 56

B. 221

E. 124

C. 123

F. 116

30) A $70 shirt now selling for $42 is discounted by what percent?

A. 45 %

C. 40 %

B. 28 %

D. 35 %

www.mathnotion.com   { 282 }

31) The score of Zoe was one third of Emma and the score of Harper was twice that of Emma. If the score of Harper was 98, what is the score of Zoe?

A. 1433

B. 16.33

C. 18

D. 17

32) Three fourth of 48 is equal to $\frac{4}{5}$ of what number?

A. 52.5

B. 45

C. 15.5

D. 20.5

33) A bag contains 20 balls: five green, three black, five blue, four red and three white. If 18 balls are removed from the bag at random, what is the probability that a blue ball has been removed?

A. $\frac{1}{4}$

B. $\frac{9}{10}$

C. $\frac{2}{18}$

D. $\frac{1}{10}$

34) What is the value of 6^3?

Write your answer in the box below.

[ ]

35) What is the median of these numbers? 2, 15, 15, 14, 9,1 9, 6, 22, 21

A. 15

B. 21

C. 9

D. 19

36) How many tiles of 8 cm² is needed to cover a floor of dimension 9 cm by 32 cm?

    A. 35   C. 36

    B. 25.5   D. 38

37) A chemical solution contains 5% alcohol. If there is 15.5 ml of alcohol, what is the volume of the solution?

    A. 410 ml   C. 420 ml

    B. 310 ml   D. 340 ml

38) The radius of the following cylinder is 6 inches and its height are 8 inches. What is the surface area of the cylinder in square inches? (π equals 3.14)

    Write your answer in the box below.

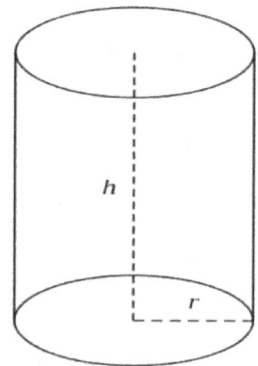

39) The average high of 14 constructions in a town is 156 m, and the average high of 6 towers in the same town is 150 m. What is the average high of all the 20 structures in that town?

    A. 154.2   C. 155.5

    B. 148   D. 145

40) The price of a laptop is decreased by 25% to $625. What is its original price?

   A. 833.33   C. 781.25

   B. 488.75   D. 361.25

41) In 1999, the average worker's income increased $2,150 per year starting from $24,000 annual salary. Which equation represents income greater than average? (I = income, x = number of years after 1999)

   A. $I > -2{,}150\,x + 24{,}000$

   B. $I > 2{,}150\,x + 24{,}000$

   C. $I < 24{,}000\,x + 2150$

   D. $I < -24{,}000\,x + 2{,}150$

42) A boat sails 24 miles south and then 18 miles east. How far is the boat from its start point?

   A. 42 miles   C. 30 miles

   B. 27 miles   D. 108 miles

43) If 96 % of F is 12 % of M, then F is what percent of M?

   A. 800 %   C. 60 %

   B. 1400 %   D. 400 %

44) Which graph corresponds to the following inequalities?

$$y \geq x - 3$$

$$3x + y \leq 6$$

A.

B.

C.

D.

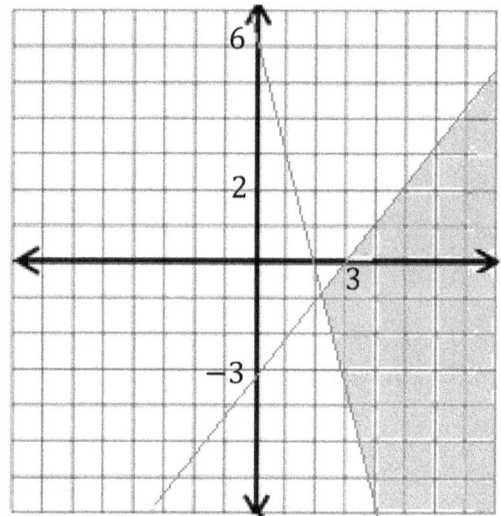

45) From last year, the price of gasoline has increased from $2.25 per gallon to

$3.5 per gallon. The new price is what percent of the original price?

A. 155.55 %

B. 165.55 %

C. 135.5 %

D. 175.5 %

46) How many possible outfit combinations come from nine shirts, six slacks, and

8 times?

Write your answer in the box below.

┌─────────────────────┐
│   │
└─────────────────────┘

End of GED Mathematical Reasoning Practice Test 2

Chapter 12 : Answers and Explanations
GED Math Practice Tests
Answer Key

❋Now, it's time to review your results to see where you went wrong and what areas you need to improve!

GED Math Practice Test

Practice Test 1							Practice Test 2						
1	A	16	B	31	D		1	B	16	24	31	B
2	C	17	B	32	B		2	A	17	A	32	B
3	B	18	C	33	C		3	D	18	C	33	A
4	−76	19	A-E	34	C		4	−21	19	A-E	34	216
5	B	20	64.04	35	93		5	B	20	A	35	A
6	B	21	C	36	D		6	C	21	B	36	C
7	729	22	D	37	B		7	−4.5	22	130	37	B
8	C	23	A–D	38	A		8	C	23	C	38	527.52
9	C	24	A	39	A		9	A	24	208	39	A
10	D	25	C	40	B		10	A	25	C-E	40	A
11	B	26	C	41	D		11	A	26	B	41	B
12	7	27	C	42	50		12	D	27	B	42	C
13	A	28	C	43	D		13	C	28	C	43	A
14	C	29	24.12	44	C		14	D	29	A-B	44	B
15	B	30	A	45	A		15	B	30		45	A
				46	$\frac{-4}{5}$						46	432

Practice Test 1
GED Mathematical Reasoning

1) Answer: A

Write a proportion and solve for the missing number.

$\frac{45}{28} = \frac{9}{x} \rightarrow 45x = 9 \times 28 = 252$

$45x = 252 \rightarrow x = \frac{252}{45} = 5.6$

2) Answer: C

Plug in the value of x and y. $-3(x + 2y) + 2(2 - x)^2$ when $x = 1$ and $y = -3$

$= -3(1 + 2(-3)) + 2(2 - 1)^2 = -3(1 - 6) + 2(2 - 1))^2 = (-3)(-5) + 2(1)^2 = 15 + 2 = 17$

3) Answer: B

The equation of a line in slope intercept form is: $y = mx + b$

Solve for y. $2x - 3y = 6 \rightarrow -3y = -2x + 6$

Divide both sides by (-3). Then: $y = \frac{2}{3}x - 2$; The slope of this line is $\frac{2}{3}$.

The product of the slopes of two perpendicular lines is -1. Therefore, the slope of a line that is

perpendicular to this line is:

$m_1 \times m_2 = -1 \Rightarrow \frac{2}{3} \times m_2 = -1 \Rightarrow m_2 = \frac{-1}{\frac{2}{3}} = -\frac{3}{2}$

4) Answer: -76

Use PEMDAS (order of operation):

$[5 \times (-14)] + ([[(-6) \times (-11)] \div 6) + (-17) = [-70] + (66 \div 11) - 17 = -70 + 11 - 17 = -76$.

5) Answer: B

To solve absolute values equations, write two equations.

$-6x + 3$ can equal positive 33, or negative 33. Therefore,

$-6x + 3 = 33 \Rightarrow -6x = 30 \Rightarrow x = -5$

$-6x + 3 = -33 \Rightarrow -6x = -33 - 3 = -36 \Rightarrow x = 6$

Find the product of solutions: $-5 \times 6 = -30$

6) Answer: B

Simplify and combine like terms.

$(2x^4 - x^2 - x^3) - (2x^2 - 3x^4 + x^3) \Rightarrow (2x^4 - x^2 - x^3) - 2x^2 + 3x^4 - x^3 \Rightarrow 5x^4 - 2x^3 - 3x^2$

7) Answer: 729

$3^6 = 3 \times 3 \times 3 \times 3 \times 3 \times 3 = 729$

8) Answer: C

Two times of 30,000 is 60,000. One six of them cancelled their tickets.

One six of 60,000 equal 10,000 ($\frac{1}{6} \times 60,000 = 10,000$).

$(60,000 - 10,000 = 50,000)$ fans are attending this week

9) Answer: C

the population is increased by 12% and 18%. 12% increase changes the population to 112% of original population.

For the second increase, multiply the result by 118%: $(1.12) \times (1.18) = 1.322 = 132.2\%$

32.2 percent of the population is increased after two years.

10) Answer: D

Solve for $x \Rightarrow 8 \leq 2x + 6 < 18 \Rightarrow$ (add -6 all sides) $8 - 6 \leq 2x + 6 - 6 < 18 - 6 \Rightarrow 2 \leq 2x < 12 \Rightarrow$ (divide all sides by 2) $1 \leq x < 6$

x is between 1 and 6 Choice D represent this inequality.

11) Answer: B

To get a sum of 3 for two dice, we can get 2 different options: $(1, 2), (2, 1)$

To get a sum of 5 for two dice, we can get 4 different options: $(1, 4), (2, 3), (4, 1), (3, 2)$

To get a sum of 7 for two dice, we can get 6 different options:

$(1, 6), (2, 5), (3, 4), (4, 3), (5, 2), (6, 1),$

Therefore, there are 12 options to get the sum of 3 or 5 or 7.

Since, we have $6 \times 6 = 36$ total options, the probability of getting a sum of 3 or 5 or 7 is 12 out of 36 or $\frac{1}{3}$.

12) Answer: 7

Use formula of rectangle prism volume.

V = (length) (width) (height) $\Rightarrow 3,276 = (36) (13)$ (height)

$\Rightarrow$ height $= 3,276 \div 468 = 7$

13) Answer: A

average (mean) = $\frac{\text{sum of terms}}{\text{number of terms}} \Rightarrow 85 = \frac{\text{sum of terms}}{25} \Rightarrow$ sum $= 85 \times 25 = 2{,}125$

The difference of 45 and 20 is 25. Therefore, 25 should be subtracted from the sum. $2{,}125 - 25 = 2{,}100$

mean = $\frac{\text{sum of terms}}{\text{number of terms}} \Rightarrow$ mean $= \frac{2{,}100}{25} = 84$

14) Answer: C

Change the numbers to decimal and then compare.

$\frac{1}{12} = 0.083;$ $0.5;$ $45\% = 0.45;$ $\frac{1}{5} = 0.2$

Therefore $\frac{1}{12} < \frac{1}{5} < 45\% < 0.5.$

15) Answer: B

Volume of a box = length $\times$ width $\times$ height $= 5 \times 8 \times 3 = 120$

16) Answer: B

average = $\frac{\text{sum of terms}}{\text{number of terms}} \Rightarrow 28 = \frac{\text{sum of 6 numbers}}{6} \Rightarrow$ sum of 8 numbers $= 28 \times 6 = 168$

$35 = \frac{\text{sum of 4 numbers}}{4} \Rightarrow$ sum of 4 numbers $= 4 \times 35 = 140$

sum of 6 numbers – sum of 4 numbers = sum of 2 numbers

$168 - 140 = 28$

average of 2 numbers $= \frac{28}{2} = 14.$

17) Answer: B

Solving Systems of Equations by Elimination

Multiply the first equation by (4), then add it to the second equation.

$\begin{array}{r} 4x + y = -2 \\ 4(-x - 2y = -3) \end{array} \Rightarrow \begin{array}{r} 4x + y = -2 \\ -4x - 8y = -12 \end{array} \Rightarrow -7y = -14 \Rightarrow y = 2$

18) Answer: C

The width of the rectangle is triple its length. Let x be the length.

Then, $width = 3x$

Perimeter of the rectangle is 2 (width + length) $= 2(3x + x) = 112$

$\Rightarrow 8x = 112 \Rightarrow x = 14;$ Length of the rectangle is 14 meters.

19) Answer: A and E

(In the stadium the ratio of home fans to visiting fans in a crowd is 5:7. Therefore, total number of fans must be divisible by 12: $5 + 7 = 12$. Let's review the choices:

A. 49,200: $49,200 \div 12 = 4100$

B. 44,600: $44,600 \div 12 = 3716.66$

C. 54,400: $54,400 \div 12 = 4533.33$

D. 51,400: $51,400 \div 12 = 4283.33$

E. 63,000: $63,000 \div 12 = 5250$

Only choices A and E when divided by 12 result a whole number.

20) Answer: 64.04

$\sqrt{256.32} = 16.01$

Four times the side of the square is the perimeter: $4 \times 16.01 = 64.04$

21) Answer: C

$\text{Probability} = \dfrac{number\ of\ desired\ outcomes}{number\ of\ total\ outcomes} = \dfrac{11}{18+21+11+12} = \dfrac{11}{62}$

22) Answer: D

Let x be the side.

Use Pythagorean Theorem: $a^2 + b^2 = c^2$

$x^2 + x^2 = 8^2 \Rightarrow 2x^2 = 64 \Rightarrow x^2 = 32 \Rightarrow x = \sqrt{32}$

The area of the square is: $\sqrt{32} \times \sqrt{32} = 32$

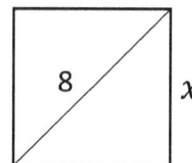

23) Answer: A and D

$4x - 2y = -4$. Plug in the values of x and y from choices provided. Then:

A. $(-1, 4): 4(-1) - 2(4) = -4 - 8 = -12$, This is NOT true!

B. $(1, 4): 4(1) - 2(4) = 4 - 8 = -4$, This is true!

C. $(-2, 2): 4(-2) - 2(2) = -8 - 4 = -12$, This is NOT true

D. $(0\ 2): 4(0) - 2(2) = 0 - 4 = -4$, This is true!

24) Answer: A

To find the number of possible outfit combinations, multiply number of options for each factor:

$3 \times 5 \times 8 = 120$

25) Answer: C

Th ratio of boy to girls is 9:11. Therefore, there are 9 boys out of 20 students. To find the answer, first divide the total number of students by 20, then multiply the result by 9.

$100 \div 20 = 5 \Rightarrow 5 \times 9 = 45$. There are 45 boys and 55 (100 − 45) girls.

So, 10 more boys should be enrolled to make the ratio 1:1

26) Answer: C

First, find the sum of six numbers.

$\text{average} = \frac{\text{sum of terms}}{\text{number of terms}} \Rightarrow 42 = \frac{\text{sum of 6 numbers}}{6} \Rightarrow$ sum of 6 numbers $= 6 \times 42 = 252$

The sum of 6 numbers is 252. If a seventh number that is greater than 54 is added to these numbers, then the sum of 7 numbers must be greater than 306 (252 + 54 = 306), then the average of these new numbers is: $\text{average} = \frac{\text{sum of terms}}{\text{number of terms}} = \frac{306}{7} = 43.71$

Since the number is bigger than 54. Then, the average of seven numbers must be greater than 43.71.

27) Answer: C

The perimeter of the trapezoid is 55 cm.

Therefore, the missing side (height) is = 55 − (15 + 12 + 14) = 14

Area of a trapezoid: $A = \frac{1}{2} h (b_1 + b_2) = \frac{1}{2} (14)(12 + 14) = 182$

28) Answer: C

The probability of choosing a Diamonds is $\frac{26}{52} = \frac{1}{2}$

29) Answer: 24.12

Let x be the width of the rectangle.

Use Pythagorean Theorem: $a^2 + b2 = c^2$

$x^2 + 4^2 = 9^2 \Rightarrow x^2 + 16 = 81 \Rightarrow x^2 = 81 - 16 = 65 \Rightarrow x = 8.06$.

Perimeter of the rectangle = 2 (length + width) = 2 (8.06 + 4) = 24.12

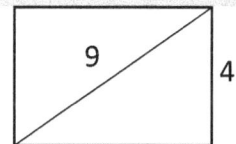

30) Answer: A

Surface Area of a cylinder = $2\pi r (r + h)$,

The radius of the cylinder is 6 (12÷ 2) inches, and its height is 6 inches. Therefore, Surface Area of a cylinder = $2\pi (6)(6 + 18) = 288\pi$

31) Answer: D

Simplify: $4x^3y^4(-xy^3)^3 = 4x^3y^4(-x^3y^9) = -4x^6y^{13}$

32) Answer: B

3,200 out of 60,800 equals to $\frac{3,200}{60,800} = \frac{1}{19}$

33) Answer: C

Write the numbers in order: 5,6,11,13,14,18,24

Median is the number in the middle. So, the median is 13.

34) Answer: C

Use simple interest formula: $I = prt$ (I = interest, p = principal, r = rate, t = time)

$I = (21000)(0.0315)(6) = 3,969$

35) Answer: 93

Mrs. Thomson needs a 90% average to pass for four exams. Therefore, the sum of 4 exams must be at lease $4 \times 90 = 360$

The sum of 3 exams is: $87 + 82 + 98 = 267$

The minimum score Mrs. Thomson can earn on her fourth and final test to pass is:

$$360 - 267 = 93$$

36) Answer: D

The distance between Daniel and Noa is 10 miles. Daniel running at 2 miles per hour and Noa is running at the speed of 4 miles per hour. Therefore, every hour the distance is 2 miles less. $10 \div 2 = 5$.

37) Answer: B

Plug in 93 for F and then solve for C.

$$C = \frac{5}{9}(F - 32) \Rightarrow C = \frac{5}{9}(93 - 32) \Rightarrow C = \frac{5}{9}(61) = 33.88$$

38) Answer: A

Let x be the number of new shoes the team can purchase. Therefore, the team can purchase $95\,x$. The team had $31,000 and spent $15,800.

Now, write the inequality: $95x + 15,800 \le 31,000$

39) Answer: A

Let x be the number. Write the equation and solve for x.

$70\% \ of\ x = 36 \Rightarrow 0.70x = 36 \Rightarrow x = 36 \div 0.70 = 51.43$

40) Answer: B

Let x be all expenses, then $\frac{22}{100}x = \$720 \to x = \frac{100 \times \$720}{22} = \$3,272.72$

He spent for his rent: $\frac{25}{100} \times \$3,272.72 = \$818.18$

41) Answer: D

The failing rate is 65 out of 175, $\frac{65}{175}$

Change the fraction to percent: $\frac{65}{175} \times 100\% = 37.14\%$

37.14 percent of students failed. Therefore, 62.86 percent of students passed the exam.

42) Answer: 50

Use the information provided in the question to draw the shape.

$a^2 + b^2 = c^2$

Use Pythagorean Theorem: $30^2 + 40^2 = c^2$

$\Rightarrow 900 + 1600 = c^2 \Rightarrow 2500 = c^2 \Rightarrow c = 50.$

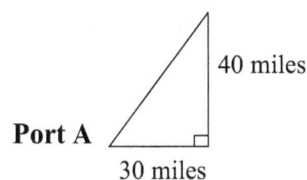

43) Answer: D

$\frac{80}{125}$, simplify by 5, then the number is the square root of $\frac{16}{25}$

$\sqrt{\frac{16}{25}} = \frac{4}{5}$; The cube of the number is: $(\frac{4}{5})^3 = \frac{64}{125}$

44) Answer: C

Isolate and solve for x. $\frac{1}{3}x + \frac{3}{7} = \frac{1}{2} \Rightarrow \frac{1}{3}x = \frac{1}{2} - \frac{3}{7} = \frac{1}{14} \Rightarrow \frac{1}{3}x = \frac{1}{14}$

Multiply both sides by the reciprocal of the coefficient of x.

$\frac{1}{3}x = \frac{1}{14} \Rightarrow x = \frac{3}{14}$

45) Answer: A

First, find the number.

Let x be the number. Write the equation and solve for x.

120 % of a number is 85, then: $1.2 \times x = 85 \Rightarrow x = 85 \div 1.2 = 70.83$

60 % of 70.83 is: $0.6 \times 70.83 = 42.49$

46) Answer: $-\frac{4}{5}$

Solve for y. $-4x - 5y = 32 \Rightarrow -5y = 32 + 4x \Rightarrow y = -\frac{4}{5}x - \frac{32}{5}$

The slope of the line is $-\frac{4}{5}$

Practice Test 2

GED Mathematical Reasoning

1) Answer: B

Plug in each pair of numbers in the equation: $x - \frac{3}{2}y = -4$

    A. $(-1, -2)$: $(-1) - \frac{3}{2}(-2) = 2$

    B. $(-1, 2)$: $(-1) - \frac{3}{2}(2) = -4$

    C. $(-2, 1)$: $(-2) - \frac{3}{2}(1) = -3.5$

    D. $(1, -2)$: $(1) - \frac{3}{2}(-2) = 7$.

2) Answer: A

To find the discount, multiply the number by (100% – rate of discount).

Therefore, for the first discount we get: (250) (100% – 15%) = (250) (0.85)

For the next 5 % discount: (250) (0.85) (0.95).

3) Answer: D

$-2x + 9 = 1.5 \rightarrow -2x = 1.5 - 9 = -7.5 \rightarrow x = \frac{-7.5}{-2} = 3.75$

Then; $3x - \frac{3}{2} = 3(3.75) - \frac{3}{2} = 11.25 - 1.5 = 9.75$.

4) Answer: −21

Use PEMDAS (order of operation):

$4 \times (-3) + 1 + 2(-8 - 9 \times 3) \div 7 = -12 + 1 + 2(-8 - 27) \div 7 = -11 + 2(-35) \div 7 =$

$-11 - 70 \div 7 = -11 - 10 = -21$.

5) Answer: B

First draw an isosceles triangle. Remember that two

sides of the triangle are equal. Let put a for the legs.

Then: $a = 6 \Rightarrow$ area of the triangle is $= \frac{1}{2}(6 \times 6) = \frac{36}{2} = 18 \ cm^2$.

Isosceles right triangle

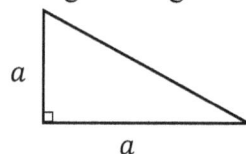

6) Answer: C

The average speed of Ryan is: 250 ÷ 5 = 50 km

The average speed of Riley is: 180 ÷ 3 = 60 km

Write the ratio and simplify. 50: 60 ⇒ 5: 6.

7) Answer: -4.5

Solving Systems of Equations by Elimination

$$\begin{aligned}3x + 4y &= -9 \\ -5x - 2y &= -6\end{aligned}$$ Multiply the first equation by 5, and second equation by 3 , then add two equations.

$$\begin{aligned}5(3x + 4y = -9) \\ 3(-5x - 2y = -6)\end{aligned} \Rightarrow \begin{aligned}15x + 20y = -45 \\ -15x - 6y = -18\end{aligned} \Rightarrow 14y = -63 \Rightarrow y = -4.5.$$

8) Answer: C

The sum of supplement angles is 180. Let x be that angle. Therefore,

$x + 6x = 180 \Rightarrow 7x = 180$, divide both sides by 7: $x = 25.71$.

9) Answer: A

Use percent formula: Part$= \frac{percent \times whole}{100}$

$490.5 = \frac{percent \times 552}{100} \Rightarrow$ (cross multiply): $490.5 = percent \times 552$

$\Rightarrow$ percent $= \frac{49,050}{552} = 88.86$

$100\% - 88.86\% = 11.14\%$.

10) Answer: A

Let x be the number. Write the equation and solve for x.

$\frac{(40-x)}{x} = 3$ (Cross multiply)

$(40 - x) = 3x$, then add x both sides. $40 = 4x$, now divide both sides by 4. $\Rightarrow x = 10$.

11) Answer: A

Use Pythagorean Theorem: $a^2 + b^2 = c^2$

$9^2 + 12^2 = C^2 \Rightarrow 81 + 144 = C^2 \Rightarrow 225 = c^2 \Rightarrow c = 15$

12) Answer: D

$\$18 \times 10 = \$180$

Petrol use: $3 \div 2 = 1.5 \, per \, hour, 10 \times 1.5 = 15$ liters

Petrol cost: $15 \times \$3.50 = \$52.5$

Money earned: $\$180 - \$52.5 = \$127.5$

13) Answer: C

If the length of the box is 40, then the width of the box is one fourth of it, 10, and the height of the box is 5 (one half of the width). The volume of the box is: $V = lwh = (40)(10)(5) = 2000$

14) Answer: D

Let x be the original price.

If the price of the sofa is decreased by 30% to $770, then: $70\% \ of \ x = 770 \Rightarrow 0.70x = 770 \Rightarrow x = 770 \div 0.70 = 1,100$

15) Answer: B

The percent of girls playing tennis is: $70\% \times 40\% = 0.70 \times 0.4 = 0.28 = 28\%$

16) Answer: 24

average$= \dfrac{sum \ of \ terms}{number \ of \ terms} \Rightarrow 25 = \dfrac{(29+27+20+x)}{4} \Rightarrow 100 = 76 + x \Rightarrow x = 24$

17) Answer: A

Use this formula: Percent of Change $= \dfrac{New \ Value - Old \ Value}{Old \ Value} \times 100\%$

$\dfrac{38,400 - 32,000}{32,000} \times 100\% = -20\%$ and $\dfrac{20,480 - 25,600}{25,600} \times 100\% = -20\%$

18) Answer: C

Let x be the smallest number. Then, these are the numbers:

$x, x + 1, x + 2$

average$= \dfrac{sum \ of \ terms}{number \ of \ terms} \Rightarrow 45 = \dfrac{x + (x+1) + (x+2)}{3}$

$\Rightarrow 45 = \dfrac{3x+3}{3} \Rightarrow 45 = x + 1 \Rightarrow x = 44$

19) Answer: A and E

(If you selected 3 choices and 2 of them are correct, then you get one point. If you answered 2 or 3 choices and one of them is correct, you receive one point. If you selected more than 3 choices, you won't get any point for this question.)

Area of the circle is less than 64π. Use the formula of areas of circles.

$$Area = \pi r^2 \Rightarrow \pi r^2 < 64\pi \Rightarrow r^2 < 64 \Rightarrow r < 8$$

Radius of the circle is less than 8. Let's put 8 for the radius. Now, use the circumference formula: Circumference $= 2\pi r = 2\pi (8) = 16\pi$

Since the radius of the circle is less than 8. Then, the circumference of the circle must be less than 16π. Online choices A and E are less than 16π

20) Answer: A

Use simple interest formula: I = prt (I = interest, p = principal, r = rate, t = time)

I = (7500)(0.025)(2) = 375

21) Answer: B

Add the first 4 numbers. $58 + 54 + 52 + 50 = 214$

To find the distance traveled in the next 4 hours, multiply the average by number of hours.

$Distance = Average \times Rate = 55 \times 4 = 220$

Add both numbers. $214 + 220 = 434$

22) Answer: 130.

Th ratio of boy to girls is 5:8. Therefore, there are 5 boys out of 13 students. To find the answer, first divide the total number of students by 13, then multiply the result by 5. $338 \div 13 = 26 \Rightarrow$ $26 \times 5 = 130$

23) Answer: C

Use distance formula: $Distance = Rate \times time$

$\Rightarrow 245 = 25 \times T$, divide both sides by 25. $\Rightarrow T = 9.8$ hours.

Change hours to minutes for the decimal part. $0.6 \ hours = 0.8 \times 60 = 48 \ minutes$.

24) The answer is 208.

The perimeter of the trapezoid is 60.

Therefore, the missing side (height) is $= 60 - (12 + 18 + 14) = 16$

Area of a trapezoid: $A = \frac{1}{2} h (b1 + b2) = \frac{1}{2} (16) (12 + 14) = 208$

25) Answer: C and E

The equation of a line is in the form of $y = mx + b$, where m is the slope of the line and b is the $y - intercept$ of the line.

Two points $(1, -1)$ and $(2,1)$ are on the line A. Therefore, the slope of the line A is:

$slope \ of \ line \ A = \frac{y_2 - y_1}{x_2 - x_1} = \frac{1-(-1)}{2-1} = \frac{2}{1} = 2$

The slope of line A is 2. Thus, the formula of the line A is:

$y = mx + b = 2x + b$, choose a point and plug in the values of x and y in the equation to solve for b. Let's choose point $(2, 1)$. Then:

$$y = 2x + b \rightarrow 1 = 4 + b \rightarrow b = 1 - 4 = -3$$

The equation of line A is: $y = 2x - 3$

Now, let's review the choices provided:

A. $(1, 5)$ $y = 2x - 3 \rightarrow 5 = 2 - 3 = -1$, This is not true.

B. $(2, 5)$ $y = 2x - 3 \rightarrow 5 = 4 - 3 = 1$, This is not true!

C. $(0, -3)$ $\quad y = 2x - 3 \rightarrow -3 = 0 - 3 = -3$, This is true.

D. $(0, -4)$ $\quad y = 2x - 3 \rightarrow -4 = 0 - 3 = -3$, This is not true!

E. $(-2, -7)$ $\quad\quad y = 2x - 3 \rightarrow -7 = -4 - 3 = -7$, This is true.

26) Answer: B

Use percent formula: $Part = \frac{percent \times whole}{100}$

$54 = \frac{percent \times 48}{100} \Rightarrow \frac{54}{1} = \frac{percent \times 48}{100}$, cross multiply.

$5400 = percent \times 48$, divide both sides by 48. $112.5 = percent$

27) Answer: B

To find the discount, multiply the number by (100% – rate of discount).

Therefore, for the first discount we get: $(100\% - 20\%)(E) = (0.80)E$

For increase of 5 %:

$(0.80)E \times (100\% + 5\%) = (0.80)(1.05) = 0.84E$.

28) Answer: C

The weight of 15 meters of this rope is: $15 \times 350g = 5250g$

1 kg = 1,000 g, therefore, $5250 \, g \div 1000 = 52.50 kg$

29) Answer: A and B

(If you selected 3 choices and 2 of them are correct, then you get one point. If you answered 2 or 3 choices and one of them is correct, you receive one point. If you selected more than 3 choices, you won't get any point for this question.)

Some of prime numbers are: 2, 3, 5, 7, 11, 13,17,19

Find the product of two consecutive prime numbers:

$5 \times 7 = 35$ (bingo!)

$7 \times 11 = 77$ (not in the options)

$11 \times 13 = 143$ (yes!)

$13 \times 17 = 221$ (not in the options)

Choices A and B are correct.

30) Answer: C

Use the formula for Percent of Change: $\frac{New\ Value - Old\ Value}{Old\ Value} \times 100 \%$

$\frac{42-70}{70} \times 100 \% = -40 \%$ (negative sign here means that the new price is less than old price).

31) Answer: B

If the score of Harper was 98, therefore the score of Emma is 49. Since, the score of Zoe was one third of Emma, therefore, the score of Zoe is 16.33.

32) Answer: B

Let x be the number. Write the equation and solve for x.

$\frac{3}{4} \times 48 = \frac{4}{5}$. $x \Rightarrow \frac{3 \times 48}{4} = \frac{4x}{5}$, use cross multiplication to solve for x, $15 \times 48 = 4x \times 4 \Rightarrow 720 = 16x \Rightarrow x = 45$

33) Answer: A

If 20 balls are removed from the bag at random, there will be five ball in the bag.

The probability of choosing a blue ball is 5 out of 20. Therefore, the probability of not choosing a white ball is 18 out of 20 and the probability of having not a white ball after removing 18 balls is the same ($\frac{5}{20} = \frac{1}{4}$).

34) Answer: 216

$6^3 = 6 \times 6 \times 6 = 216$

35) Answer: A

Write the numbers in order: 2, 6, 9, 14, 15, 15, 19, 21, 22

Since we have 9 numbers (9 is odd), then the median is the number in the middle, which is 15.

36) Answer: C

The area of the floor is: $9 \, cm \times 32 \, cm = 288 \, cm^2$

The number of tiles needed $= 288 \div 8 = 36$

37) Answer: B

5% of the volume of the solution is alcohol. Let x be the volume of the solution. Then: 5% of $x = 15.5$ ml

$0.05x = 15.5 \Rightarrow \frac{5x}{100} = \frac{155}{10}$ cross multiply, $50x = 15500 \Rightarrow$ (devide by 50) $x = 310$

38) Answer: 527.52.

Surface Area of a cylinder = $2\pi r \, (r + h)$,

The radius of the cylinder is 6 inches, and its height are 8 inches. π is 3.14. Then: Surface Area of a cylinder = $2 \, (3.14) \, (6) \, (6 + 8) = 527.52$

39) Answer: A

$average = \frac{sum \, of \, terms}{number \, of \, terms}$

The sum of the high of all constructions is: $14 \times 156 = 2,184$ m

The sum of the high of all towers is: $6 \times 150 = 900$ m

The sum of the high of all building is: $2,184 + 900 = 3,084$

$\text{average} = \dfrac{3,084}{20} = 154.2$

40) Answer: A

Let x be the original price.

If the price of a laptop is decreased by 25% to $625, then:

$75\% \ of \ x = 625 \Rightarrow 0.75x = 625 \Rightarrow x = 625 \div 0.75 = 833.33$

41) Answer: B

Let x be the number of years. Therefore, $2,150 per year equals $2,150x$.

starting from $24,000 annual salary means you should add that amount to $2,150x$.

Income more than that is: $I > 2,150x + 24,000$

42) Answer: C

Use the information provided in the question to draw the shape.

Use Pythagorean Theorem: $a^2 + b^2 = c^2$

$24^2 + 18^2 = c^2 \Rightarrow 576 + 324 = c^2 \Rightarrow 900 = c^2$

$\Rightarrow c = 30.$

24

18 miles

43) Answer: A

Write the equation and solve for M:

0.96 F = 0.12 M, divide both sides by 0.12, then: 0.96/0.12 F = M, therefore:

M = 8 F, and M is 8 times of F or it's 800% of F.

44) Answer: B

For each option, choose a point in the solution part and check it on both inequalities.

$$y \geq x - 3$$

$$3x + y \leq 6$$

A. Point $(0, -4)$ is in the solution section. Let's check the point in both inequalities. $-4 \geq 0$

    -3,   That's not true

    $3(0) + (-4) \leq 6 \Rightarrow -4 \leq 6,$ That's true

B. Let's choose this point $(0, 0)$; $0 \geq 0 - 3$, That's true

    $3(0) + (0) \leq 6$, That's true!

C. Let's choose this point (3, 2); $3(3) + 2 \leq 6$, That's not true!

D. Let's choose this point (5, –1); $-1 \geq 5- 3$, That's not true!

45) Answer: A

The question is this: 3.5 is what percent of 2.25?

Use percent formula: $\text{part} = \frac{percent}{100} \times whole$

$3.5 = \frac{percent}{100} \times 2.25 \Rightarrow 3.5 = \frac{percent \times 2.25}{100} \Rightarrow 350 = percent \times 2.25 \Rightarrow percent = \frac{350}{2.25} = 155.55.$

46) Answer: 432.

To find the number of possible outfit combinations, multiply number of options for each factor:

$9 \times 6 \times 8 = 432$

"End"